URBAN POLITICS

SAGE has been part of the global academic community since 1965, supporting high quality research and learning that transforms society and our understanding of individuals, groups and cultures. SAGE is the independent, innovative, natural home for authors, editors and societies who share our commitment and passion for the social sciences.

Find out more at: **www.sagepublications.com**

Edited by
Mark Davidson & Deborah Martin

URBAN POLITICS
Critical Approaches

Los Angeles | London | New Delhi
Singapore | Washington DC

Los Angeles | London | New Delhi
Singapore | Washington DC

SAGE Publications Ltd
1 Oliver's Yard
55 City Road
London EC1Y 1SP

SAGE Publications Inc.
2455 Teller Road
Thousand Oaks, California 91320

SAGE Publications India Pvt Ltd
B 1/I 1 Mohan Cooperative Industrial Area
Mathura RoadA
New Delhi 110 044

SAGE Publications Asia-Pacific Pte Ltd
3 Church Street
#10-04 Samsung Hub
Singapore 049483

Editor: Robert Rojek
Editorial assistant: Keri Dickens
Production editor: Katherine Haw
Copyeditor: Neil Dowden
Proofreader: Rosemary Morlin
Indexer: Mark Davidson and Deborah Martin
Marketing manager: Michael Ainsley
Cover design: Francis Kenney
Typeset by: C&M Digitals (P) Ltd, Chennai, India
Printed by: Replika Press Pvt. Ltd, India

Editorial arrangement and Chapters 1 and 13 © Mark Davidson
 and Deborah Martin 2014
Chapter 2 © Katherine Hankins and Deborah Martin 2014
Chapter 3 © Kevin Ward 2014
Chapter 4 © Elvin Wyly and Kathe Newman 2014
Chapter 5 © Kurt Iveson 2014
Chapter 6 © Donald McNeill 2014
Chapter 7 © John Carr 2014
Chapter 8 © Natalie Oswin 2014
Chapter 9 © Jamie Winders 2014
Chapter 10 © Susan Hanson 2014
Chapter 11 © Mark Davidson 2014
Chapter 12 © Matthew Huber 2014

First published 2014

Apart from any fair dealing for the purposes of research or private study, or criticism or review, as permitted under the Copyright, Designs and Patents Act, 1988, this publication may be reproduced, stored or transmitted in any form, or by any means, only with the prior permission in writing of the publishers, or in the case of reprographic reproduction, in accordance with the terms of licences issued by the Copyright Licensing Agency. Enquiries concerning reproduction outside those terms should be sent to the publishers.

Library of Congress Control Number: 2013936854

British Library Cataloguing in Publication data

A catalogue record for this book is available from the British Library

ISBN 978-0-85702-397-1
ISBN 978-0-85702-398-8 (pbk)

CONTENTS

List of Figures — vii
List of Tables — viii
List of Boxes — ix
Notes on Contributors — x
Acknowledgements — xv

1 Thinking critically about urban politics — 1
  Mark Davidson and Deborah Martin

SECTION 1: CITY AS SETTING — 15

  Introduction to City as Setting — 17
  Mark Davidson and Deborah Martin

2 The neighbourhood as a place for (urban) politics — 23
  Katherine Hankins and Deborah Martin

3 Splintered governance: urban politics in the twenty-first century — 42
  Kevin Ward

4 Ruthless: the foreclosure of American urban politics — 55
  Elvin Wyly and Kathe Newman

SECTION 2: CITY AS MEDIUM — 77

  Introduction to City as Medium — 79
  Mark Davidson and Deborah Martin

5 Policing the city — 85
  Kurt Iveson

6 Mayors and the representation of urban politics — 100
  Donald McNeill

7   Making urban politics go away: the role of legally mandated
    planning processes in occluding city power   112
    John Carr

SECTION 3: CITY AS COMMUNITY   **131**

   Introduction to City as Community   133
   Deborah Martin and Mark Davidson

8   Queering the city: sexual citizenship in creative city Singapore   139
    Natalie Oswin

9   Making space in the multicultural city: immigrant settlement,
    neighbourhoods and urban politics   156
    Jamie Winders

10 The embedded politics of entrepreneurs   172
    Susan Hanson

11 Is class relevant to urban politics?   189
    Mark Davidson

12 The urban imaginary of nature: cities in American
    environmental politics   206
    Matthew Huber

13 Urban politics as parallax   223
    Deborah Martin and Mark Davidson

Index   231

FIGURES

2.1	Thanksgiving Day on Cheryl's front porch, Booker T. Washington neighbourhood, 11 November 2010. Cheryl has been a strategic neighbour for four years in Atlanta	33
2.2	Community garden in English Avenue, west Atlanta, 11 October 2009. This community garden was developed by residents and strategic neighbours	35
2.3	Historic home in South Atlanta, two doors down from a strategic neighbour's residence. South Atlanta, a neighbourhood of approximately 550 homes, has over two dozen strategic neighbours	36
4.1	LISA indices for 2004, 2006, 2007 and 2008:	
4.1a	Rapid defaults, surrounded by rapid defaults	64
4.1b	Rapid defaults, surrounded by long survivors	64
4.1c	Long survivors, surrounded by long survivors	64
4.1d	Long survivors, surrounded by rapid defaults	65
4.1e	No significant spatial correlation	65
5.1	A sign prohibiting graffiti, covered in and surrounded by graffiti, Sydney	89
5.2	The infamous Banksy piece which partially covered a 25-year-old piece by Robbo in London	95
7.1	Ballard Commons Park with the New Ballard Skatepark	117
7.2	Seattle's 2007 Master Plan for Skateparks	121
7.3	Opening day for Ballard Commons Park, with the skatepark in the background	124
8.1	Pink Dot formation, Singapore	145
10.1	Density of woman-owned businesses/area per census tract, Worcester City and metro area	185
10.2	A virtuous circle? How place rootedness can lead to change	186
11.1	Bank Junction, City of London: a key site of global financial networks	196

TABLES

4.1	Default speeds in Essex County, New Jersey	61
4.2	Model fit for borrower credit history instrument	66
4.3	Countywide subprime segmentation models	68
4.4	Contrasts between quick-default and long-survivor clusters	71
9.1	The multicultural city	165
10.1	Volunteer activities linked specifically to business: Worcester entrepreneurs	182
10.2	Volunteer activities linked specifically to business: Colorado Springs entrepreneurs	183
11.1	Operational categories of the National Statistics Socio-economic Classification (NS-SEC) linked to socio-economic groups	200
12.1	The changing nature of American environmental politics	207

BOXES

1.1	Geography and politics	4
S1	The urban citizen in historical perspective	19
2.1	Strategic neighbour Crissy Brooks describes goals	30
2.2	Strategic neighbour Scott Dewey describes dialogue as a goal of strategic neighbouring	30
2.3	Strategic neighbour Ashlee Starr describes her Atlanta neighbourhood	30
S2	Foucault and the mediums of social control	81
6.1	Mayors and social media	104
7.1	Critical legal studies and urban geography	114
8.1	Richard Florida's creative class thesis	145
11.1	Commentary on London's energy policy	196
12.1	The changing nature of American environmental politics	211
12.2	Urban political ecology and the metabolic city	212
13.1	Dilemmas of citizen–state engagements in urban politics	227

NOTES ON CONTRIBUTORS

John Carr is an Assistant Professor in the Department of Geography and Environmental Studies at the University of New Mexico. He received his PhD in Geography at the University of Washington, and his JD at the University of Texas School of Law. His research focuses on the politics of urban public space, with a focus on law and urban planning processes.

Mark Davidson is an Assistant Professor in the Graduate School of Geography at Clark University, Worcester, Massachusetts. He is an urban geographer whose research interests lie in three core areas: 'gentrification', 'urban policy, society and community' and 'metropolitan development, planning and architecture'. His research is international in scope, including work in Europe, North America and Australia. He has authored and co-authored papers in journals such as *Environment and Planning A*, *Ethics, Place and Environment*, *Transactions of the Institute of British Geographers* and *Urban Studies*. His current research includes a continued examination of new-build gentrification, a theoretical exploration of gentrification-related displacement and the empirically informed consideration of sustainability as a key policy concept. He has held fellowships at the Nelson A. Rockefeller Centre for Public Policy and Social Science, Dartmouth College, and the Urban Research Centre, University of Western Sydney. He holds a BA (Hons) and PhD in Geography from King's College London.

Katherine Hankins is an Associate Professor of Geography in the Department of Geosciences at Georgia State University. She is an urban geographer, whose expertise is in urban politics, neighbourhood activism and community development. Her current research examines the conceptions of social and spatial justice in the discourses and practices of individuals and organizations, whose work is motivated by principles of Christian community development. In particular, she is interested in the politics of 'strategic neighbouring' in inner-city neighbourhoods, and the subjectivities and spaces produced by efforts associated with 'gentrification with justice.' This work builds on her recent research on the social and political dynamics of charter-school activism. Her work has been published in *Urban Studies*, *International Journal of Urban and Regional Research*, *Antipode*, *Urban Geography*, *Journal of Urban History* and *Social* *and Cultural Geography*, among others. She is currently collaborating with an interdisciplinary team of scholars and community partners to address social and environmental disparities in Atlanta neighbourhoods. She holds a Master's in geography from the University of Arizona and a PhD in geography from the University of Georgia.

Susan Hanson is a Professor of Geography (Emerita) at Clark University. She is an urban geographer with interests in gender and economy, transportation, local labour markets and sustainability. Her research has focused on the relationship between the urban built environment and people's everyday mobility within cities; within this context, questions of access to opportunity, and how gender affects access, have been paramount. Her chapter in this book grew out of a larger study focused on understanding how gender, geographic opportunity structures and geographic rootedness shape entrepreneurship in cities. She has been the editor of several academic journals including *The Annals of the Association of American Geographers*, *Urban Geography* and *Economic Geography*. Her publications include the books *Ten Geographic Ideas that Changed the World* (Rutgers University Press 1997), *Gender, Work, and Space* (with Geraldine Pratt) (Routledge 1995), *The Geography of Urban Transportation* (Guilford Press 1986), and numerous journal articles and book chapters. A past president of the Association of American Geographers, Hanson has served on many national and international committees in geography, transportation and the social sciences. Her BA is from Middlebury College, and before earning her PhD at Northwestern University she was a Peace Corps Volunteer in Kenya.

Matthew Huber is an Assistant Professor of Geography at Syracuse University. He earned his PhD in Geography from Clark University in 2009. Professor Huber's research focuses on energy, oil, mining and the political economy of capitalism. His recently published book from the University of Minnesota Press is entitled *Lifeblood: Oil, Freedom, and the Forces of Capital*. His work cuts across human geography in examining the economic, political and urban aspects of nature–society relationships. He has published articles in *Urban Geography*, *The Annals of the Association of American Geographers*, *Antipode* and *Geoforum*.

Kurt Iveson is Senior Lecturer in Urban Geography and teaches at the University of Sydney. His research is guided by a desire for social and spatial justice, and focuses especially on the relationship between cities, publics and politics. He is the author of *Publics and the City* (Wiley-Blackwell, 2007) and co-author of *Planning and Diversity in the City: Redistribution, Recognition and Encounter* (Palgrave, 2008). He also maintains the blog Cities and Citizenship (http://citiesandcitizenship.blogspot.com).

Deborah Martin is Associate Professor in the Graduate School of Geography at Clark University. She is an urban geographer with interests in urban politics, law, place identity, place-making and qualitative methods. Her current research projects include investigation of the legal dynamics of group home siting in Massachusetts and New York; the place-identity and policy impacts of the Asian Longhorned Beetle infestation in Worcester, Massachusetts; and dynamics and implications of 'local' governance in urban policy. She has published in journals such as *Annals of the Association of American Geographers, Antipode, Gender Place and Culture, Environment and Planning A, International Journal of Urban and Regional Research, Progress in Human Geography, Transactions of the Institute of British Geographers* and *Urban Geography*. She holds a PhD in Geography from the University of Minnesota.

Donald McNeill is Professor of Urban and Cultural Geography, joining the Institute for Culture and Society at the University of Western Sydney in 2011, having previously held positions at the Urban Research Centre at UWS, King's College London, Southampton and Strathclyde. He is a recipient of an Australian Research Council Future Fellowship, in the field of 'Governing Digital Cities', running from 2012 to 2016. His work is located at the intersection of human geography, economic sociology, spatial planning, and urban design and architecture, with a particular interest in the political and cultural economy of globalization and cities. He is currently working with colleagues on Australian Research Council grants about the future of Sydney's Chinatown, and on the social history of air-conditioning in Southeast Asia, and has recently completed an ARC Discovery project, 'The production and contestation of airport territory' (2008–2010). He has published widely in human geography and urban studies, and his books include *The Global Architect: Firms, Fame and Urban Form* (Routledge, 2008), *New Europe: Imagined Spaces* (Arnold, 2004), and *Urban Change and the European Left: Tales from the New Barcelona* (Routledge, 1999).

Kathe Newman is an Associate Professor in the Urban Planning and Policy Development Program at the Edward J. Bloustein School of Planning and Public Policy and Director of the Ralph W. Voorhees Center for Civic Engagement. She holds a PhD in Political Science from the Graduate School and University Center at the City University of New York. Her research explores urban change, what it is, why it happens and what it means. Her research has explored gentrification, foreclosure, urban redevelopment and community participation. Dr Newman has published articles in *Urban Studies, International Journal of Urban and Regional Research, Urban Affairs Review, Shelterforce, Progress in Human Geography, Housing Studies, GeoJournal* and *Environment and Planning A*.

Natalie Oswin is an Assistant Professor in the Department of Geography at McGill University. She has published articles on South Africa's post-apartheid gay and lesbian movement, the cultural politics of heteronormativity in Singapore and conceptual pieces on queer geographies in such journals as *Gender, Place and Culture*, *Environment and Planning A*, *Progress in Human Geography*, *Signs: A Journal of Women in Culture and Society*, *Social and Cultural Geography* and *Transactions of the Institute of British Geographers*. She has also co-edited special issues of the journals *Environment and Planning D: Society and Space* and *Mobilities* on the themes 'Governing Intimacy' and 'Mobile City Singapore', respectively, and is co-editor of the journal *Environment and Planning D: Society and Space*.

Kevin Ward is Professor of Human Geography and Director of cities@manchester, University of Manchester (www.cities.manchester.ac.uk). His research interests are in comparative urbanism, state restructuring, policy mobilities, and urban and regional development. His current work explores urban policies to see where they have come from, the paths they have travelled and the processes of translation they have undergone. This programme of work speaks to the various financial models currently circulating to fund infrastructure in the context of the ongoing financial crisis. It is interested in the various actors and intermediaries that play a role in constructing and constituting the circuits, networks and webs in and through which 'models' are assembled, circulated, deemed to have failed, redirected, resisted and so on. He has edited and written numerous books and his work has been published in journals such as the *Annals of the Association of American Geographers*, *Antipode*, *Area*, *Environment & Planning A*, *Environment & Planning C*, *Geoforum*, *IJURR*, *Transactions of the Institute of British Geographers*, *Urban Geography* and *Urban Studies*.

Jamie Winders is Associate Professor of Geography at Syracuse University. Her research focuses on immigration, racial politics and formations, and urban and social geography. Much of her work has examined the emergence of new immigrant destinations in the United States and the changing racial and cultural politics associated with this new geography of immigrant settlement. She is the author of *Nashville in the New Millennium: Immigrant Settlement, Urban Transformation, and Social Belonging* (Russell Sage, 2013). Her work is largely ethnographic and spans geography and other disciplines including Latino studies, southern studies and sociology. Her research has also examined the social movements, political discourses and legislative trends associated with immigrant settlement in new destinations, as well as the role of social reproduction and place-making in the politics of immigration in publications with Barbara Ellen Smith. Beyond her work on immigration, Winders has published on postcolonial theory, critical pedagogy, travel writing, civil society and critical race theory. She is also co-editor of the *New Companion to Cultural Geography* (Blackwell, 2013), with Nuala Johnson and Richard Schein.

Elvin Wyly is Associate Professor of Geography and Chair of the Urban Studies Coordinating Committee at the University of British Columbia, Vancouver (www.geog.ubc.ca/~ewyly). His teaching and research focus on the spatial constitution of urban social inequality, housing markets and neighbourhood change, public policy, and the history and present condition of strategic positivism.

ACKNOWLEDGEMENTS

We would like to thank the contributors to this volume, all of whom generously devoted their valuable time in writing the chapters and responding to our feedback and criticisms, as well as that of the blind referees. We would also like to thank our editor at SAGE, Robert Rojek, for his encouragement of the project and his faith that we would get it finished. Alana Clogan and Sarah-Jayne Boyd at SAGE also provided much need support in making the project happen. The advice of our close friends and colleagues throughout the process of putting the collection together has been crucial, including Kate Boyer, Ben Gallan, Chris Gibson, Kal Gulson, Jennie Middleton, Jim Murphy, Joe Pierce and Tom Slater. We also need to thank all those friends and colleagues whom we asked to referee the chapters. Their feedback was absolutely critical in improving the collection and pushing us to bring together our diversity of contributors. Finally, we'd like to thank our respective families for providing the love and encouragement that makes these undertakings possible.

Deborah Martin and Mark Davidson

1
THINKING CRITICALLY ABOUT URBAN POLITICS

Mark Davidson and Deborah Martin

It is now trite to call into question what we mean by the terms 'urban' or 'politics'. The urban politics literature is full of attempts to develop more cohesive and/or distinctive takes on the conjunction. Theorizing urban politics is, of course, a useful exercise that mediates those countless passing deployments of the term. Indeed this mediation is a crucial political task itself. As the chapters in this collection demonstrate, there are important stakes involved in how we understand urban politics. For example, if we understand urban politics as something that happens *within cities* and across all *contestations* then we will arrive at different conclusions with regards to things such as the vibrancy of political life and democratic process.

The task of defining 'urban politics' is not only analytical, but deeply political. When we set out a definition of urban politics we identify our object of analysis. Critical theorist Slavoj Žižek (2006) refers to this general process of identification as an act of bracketing: '*the bracketing itself produces it objects*' (56; emphasis in original). Urban politics does not simply appear to us when we decide to investigate it. Rather we have to actively construct an image of urban politics to start interrogating them. We do this by abstracting from the indeterminate set of processes that constitute the city and urban life. Without accepting this task we would be faced with a vast collection of phenomena that, if we did not select a few to focus upon, would certainly leave us paralysed, unable to start the process of making sense.

So bracketing is a necessary task in order to gain knowledge. But as Žižek (2006) explains, this task is not one that can claim to be neutral: 'This bracketing is not only epistemological, it concerns what Marx called "real abstraction"' (56). What this means is that when we define our object of study, we have the concomitant task of deciding what to include and what to leave out. If we focus on city hall as the venue for urban politics, we might be leaving out important parts of any city's political fabric. Or if we concentrate explicitly on the economic drivers of urban politics, we might omit other types of social struggles. The conundrum here is the fact we must necessarily bracket. We cannot wish to capture the complete complexities of any social arrangement. So what to do?

One response might be to say that all perspectives on urban politics – or anything else for that matter – are equally valid. Or one might say that contrary perspectives must be brought into agreement; perhaps using something akin to Hegel's dialectical method. Or, as Žižek's (2006) theory of parallax view suggests, we might sometimes accept the incongruity of two perceptions and attempt to keep both in mind at the same time. Žižek equates this to the famous optical trick of 'two faces or a vase' where you either see two faces or a vase but never both. You know both exist, but nevertheless you must choose to view only one. This is how Žižek (2006) describes the parallax in philosophical terms: '... subject and object are inherently "mediated", so that an "epistemological" shift in the subject's point of view always reflects an "ontological" shift in the object itself' (17). What this means is that different perspectives – what Žižek calls 'bracketing' – on urban politics are related, even if they remain incompatible. 'Urban politics' (the object) appears as it does because we (the subject) theorize it in certain ways.

There are many consequences to this philosophical position. Here we want to develop just one, that of the necessity to see the implications of a shift in perspective. The notion of parallax perspectives makes us think about what happens in that non-space between one viewpoint and another. For example, we might decide to approach urban politics via its official state institutions; there are good reasons to do so, since this perspective can illuminate a great deal of social struggle and order. Alternatively we might examine urban politics through the idea that everyone in the city is a political actor and, consequently, urban politics is about the entirety of social relations running through the metropolis. Again there might be very good reasons for doing this. The point, however, is that we must think about what this shift in perspective means: What changes? What gets lost when we rethink the object? What types of politics and social changes can be justified from certain viewpoints? The various approaches to urban politics in this book open some political possibilities and close others. So which possibilities should remain open and what others can, perhaps, become closed? The way we theorize the city itself has a lot to do with answering these questions.

Urban Politics and the Geographies of the City

In the past decade there has been a significant rethinking of the geography of the city, and, by extension, urban politics. At risk of oversimplification, this rethinking has involved a shift from reading the city as a discrete space with its own internal politics to a more relational view of urbanism. A good example of the former is provided by John (2009: 17): 'At its most straightforward, urban politics is about authoritative decision-making at a smaller scale than national units ... the focus of interest is at the sub-national level with particular reference to the political actors and institutions operating there.' Within this framing, urban politics is contained within cities. These internal politics can then be related to smaller (i.e. neighbourhood) and larger (i.e. national) scales. The nature of this containment has been conceptualized differently. Some have suggested the collective consumption issues (i.e. schooling, transit) necessarily generate localized political communities (Castells, 1977; Saunders, 1981). Others, particularly those interested in urban history, have pointed towards the political consciousness that arose when people started living within cities (Nash, 1979).

Within the urban politics literature, jurisdictional boundaries of cities and municipalities have often served as a foundation for those who theorize the urban as a bounded space (Logan and Molotch, 1987; Taylor, 2004). This perspective has often motivated studies that look for the particular combination of factors within an urban environment, such as city-size and demographics, which help explain variables such as citizen participation (Oliver, 2000). Those interested in multi-level governance (Hooghe and Marks, 2003) have also adopted this approach to conceptualize the city, viewing it as one part of a set of nested scales (i.e. neighbourhood, city, region, nation, global); the focus of analysis being the identification of the particular nature of governance and development within the city and how this relates to other scales. This tradition of urban studies continues, particularly in the context of what many see as a reworking of scale relations (Swyngedouw, 1997; Brenner, 2001).

A bounded city perspective has become increasingly problematized with a growing recognition that socio-spatial relations have transformed as a consequence of globalization processes (Marcuse and van Kempen, 2000). Indeed, in the context of entrepreneurial governance, mobile capital and communication technologies, scholarly views of the urban system have transformed significantly, in part as a result of thinking about the production of space as being multi-faceted and simultaneous, irrespective of national and municipal boundaries (Harvey, 1989; Lefèbvre 1991). A major consequence is that proximity and nation-state relations are now less relevant to urban governance (Swyngedouw, 1997). This means that if we are to understand the processes occurring within particular city spaces, we need to be cognizant of the varied set of relations they maintain as opposed to taking for granted that, for example, neighbouring municipalities and national governments will be the most significant.

The shift away from a bounded conceptualization of urban politics has been part of a more general critique of scalar perspectives. For Marston et al. (2005), a move away from scalar perspectives should be total, preferring instead to understand relations as networked; or to use Leitner's (2004) theorization, to move from a concern from vertical relations to horizontal relations. No longer, it is argued, can we therefore see the world in a three-scale structure (see Taylor, 1982): micro (urban), meso (nation) and macro (global). Whilst the view that scale is now redundant as a theoretical tool is not held here, the emphasis on horizontal geographies (networks; non-distanced relations; flows) is recognized as important to understanding the constitution of the city. We therefore require an understanding of the city that avoids the search for some essential spatiality. Rather we need an approach that captures the multiplicity of socio-spatialities and engages in a dialogue about the relative epistemologies developed using different perspectives on the urban (Massey, 2007).

As part of a wider attempt to rethink the geographies of the city, many now stress the relational nature of urbanism, building on foundational ideas of urbanization as a process, but incorporating more explicitly relations of agents (individuals and institutions) as well as capital (Pierce et al., 2011). Within the context of globalization processes, the idea of a discretely bounded and/or scalar political community has been either abandoned or supplemented by reading cities as inter-connected and inter-constituted.

> ### BOX 1.1   GEOGRAPHY AND POLITICS
>
> Debates about how to understand the relationship between space and politics have been developing quickly in recent years. Often-used references to 'cities' or 'nations' as scalar entities have been widely critiqued as scholars working in various fields have attempted to shift away from purely scalar-based theories. As a result you will now see much discussion of networks, topology and assemblages in the geographical literature. An example of influential work in this area is John Allen and Allan Cochrane's (2010) retheorizing of state power, where they see it as 'multiple, overlapping, tangled, interpenetrating, as well as relational' (1087). The city or nation-state is therefore seen not as a discrete domain of power, but rather as a set of topological arrangements that exert influence in a variety of diverse spaces. Studying urban or national politics therefore becomes concerned with tracing out the contours of power that political actors create over time and space. Put simply, we cannot rely on the city or nation scale as a guide to know where politics happen.

> **Suggested further reading**
>
> **Critiquing scale-based approaches**
>
> Marston, S.A., Jones, III, J.P. and Woodward, K. 2005. 'Human geography without scale', *Transactions of the Institute of British Geographers*, 30(4): 416–432.
> Commentary: Leitner, H. and Miller, B. (2007) 'Scale and the limitations of ontological debate: a commentary on Marston, Jones and Woodward', *Transactions of the Institute of British Geographers*, 32(1): 116–125.
>
> **Thinking space relationally**
>
> Allen, John (2003) *Lost Geographies of Power*. Blackwell: Oxford.
> Featherstone, David (2008) *Resistance, Space and Political Identities: The Making of Counter-Global Networks*. Oxford: Wiley-Blackwell.
> Jacobs, J. (2011) 'Urban geographies I: still thinking cities relationally', *Progress in Human Geography*, 38(3): 412–422.
>
> **City as an assemblage**
>
> McFarlane, C. (2011) 'Assemblage and critical urbanism', *City: Analysis of Urban Trends, Culture, Theory, Policy, Action*, 15(2): 204–224.
> Commentary: Brenner, N., Madden, D. and Wachsmuth, D. (2011) 'Assemblage urbanism and the challenges of critical urban theory', *City: Analysis of Urban Trends, Culture, Theory, Policy, Action*, 15(2): 225–240.

Some have rejected the idea of the bounded urban political community because they view it as politically regressive, closing off a recognition of the linkages among different oppressions, be they racial, gendered, or spatially based marginalization. A notable example of this line of thinking came from David Harvey (1996) when he wrote against the 'militant particularism' that he saw characterizing localized political movements. He argued that '[T]he potentiality for militant particularism embedded in place runs the risk of sliding back into a parochialist politics' (324). Harvey's rejection comes from seeing the now globally coordinated production and consumption of commodities creating geographically complex social relations. The idea that a localized, place-based political movement might transcend its own particular interests and politicize these relations just seems impossible for Harvey.

Others have taken a different position. Doreen Massey (1991, 2007) has written extensively on the politics of place in an era of economic globalization. In her work she has rejected the dualistic framing of the local and global that Harvey (1987) uses: 'The global is just as concrete as is the local place. If space is really to be thought relationally then it is no more than the sum of our relations and interconnections, and the lack of them' (Massey, 2005: 184). Fitting the world into the categories of local and global are rejected because the division obfuscates their mutual constitution. This argument has led many scholars to look towards those connections and relations that local places constitute between themselves (e.g. Featherstone, 2008).

One response to this spatially extensive relational approach might be: What about the government? Whilst different places might have become more connected in recent times, city governments still exist! Many scholars adopting the relational view of urban politics have attempted to respond to this type of question. For some this effort has involved a rethinking of state relations and the role that city governments play in constituting economic relations and enacting policy (e.g. Brenner, 2004). Others have attempted to conceptualize the changing form and operation of state power. John Allen (2004) has argued that state power does not operate within scales. He rejects the idea that a city's political authority is wielded purely within certain jurisdictional boundaries. Instead Allen chooses to understand state power as a topological arrangement: 'as a *relational effect* of social interaction where there are no pre-defined distances or simple proximities to speak of' (2004: 19; emphasis in original). Politics is therefore to be found across the multiplicity of networked relations that (re) make the city: 'the mediated relationships of power multiply the possibilities for political intervention at different moments and within a number of institutional settings' (ibid.: 29). We cannot predict, therefore, where power will connect various decision-makers, and it may be that far-flung relations define or circumscribe the choices of seemingly 'local' actors.

A topological account of the city therefore transcends the local/global by viewing state power as 'multiple, overlapping, tangled, interpenetrating, as well as relational' (Allen and Cochrane, 2010: 1087). An important consequence of taking this viewpoint is that we often have to look for state power in places that we would not associate with 'government' – a point made (and much debated) when Floyd Hunter (1953) and Clarence Stone (1989) demonstrated the wide-ranging, socially powerful politics operating through corporate and community elites in Atlanta, Georgia. Given that a city can have various forms of networked relations it may exercise its power in a potentially endless list of places that may be nested, overlapping and disparate. If we then think about urban politics in terms of contestation and struggle, our venues for such activities might need to be sorted out. We might have to reject the idea that political power within cities resides in city hall and, instead, look towards those points in time and space where state power is wielded.

Recently these relational approaches have been used to rethink urban policies. In particular, the idea that cities have their own policy-making procedures and resultant policies has been challenged, with attention being directed to the international networks of policy generation and exchange. Rather than see a city like London or New York as generating its own approaches to policy, such governing is complicated by the fact that the constitution of a city's policies occurs as a dialogue both within and outside the city and its officials. Recognizing the mobility of policy-making transforms conceptualizations of the geography of policies: 'it moves beyond the limits of the traditional political science-dominated policy transfer literature, acknowledging its insights while also arguing for a broadening of our understanding of agents of transference, a reconceiving of the sociospatial elements of how policies are made mobile, and a departure from the methodological nationalism' (Ward and McCann, 2011: 167). Along with viewing urban politics as constructed through sets of extra–local relations, we might also look at urban policies as emerging across and through cities in very particular ways. This might be the advent of a business development scheme in Chicago that makes its way to Manchester, or the organic origination of a policy choice in California that gets packaged and codified to be deployed in Pennsylvania. Whatever the particular case, we are again made cognizant of the fact that our theorization of the city and policies therein will itself shape what it is we are studying.

A challenge facing the urban politics literature is therefore to decide whether its various theorizations of the city are compatible with one another. Is there a way that the relational view of cities (and policy-making) could complement or co-exist with the idea that the city is a contained space with its own politics and policy-making? Could we transcend the current distinction between a contained space versus flexible, contingent and changing space with another perspective? Or do we need to view each perspective as incompatible and, consequently, consider the movement between the two a parallax shift? If so we would need to ask what gets lost and what is found when we see the city in either frame, as each, like a parallax view, offers insights, but obscures those of the other.

The Politics of Naming 'Politics'

But what of politics? Do we face the same types of questions with regards to the *politics* in urban politics? To some degree we do. There are clearly different conceptualizations of politics available. Politics can be associated with government and institutions. They can be examined in terms of the multitude of governance procedures that produce order. Politics can be viewed as always present in all of our actions. Or politics can be made something quite specific that goes beyond power and interaction to reordering and change (Rancière, 1999). Within this collection you will find various derivations of these approaches.

There are some common distinctions within the urban politics literature with respect to how politics are theorized. As Davies and Imbroscio (2009: 3) commented, politics is often divided into questions of government/institutions and governance. The former is quite obvious. It sees politics as occurring within and/or around those institutions that are given the power to govern. So you might be concerned with the election of city councils, the reform of institutional structures or the geography of voting patterns. Questions of governance, by contrast, tend to revolve around a more diverse array of concerns. Anne Mette Kjaer (2009) describes the relationship between these two approaches:

> Over the last two decades the term 'governance' was applied to denote a change in public administration from a set-up focusing on hierarchy and clear demarcation lines between politics and administration, and between the state and society, to an organisational set-up emphasising networks and the overlapping roles of politicians and administrators as well as of state and society actors. (137)

The spilling over of institutional regimes into society generally signals to the debt which many writings on urban politics owe Michel Foucault's (1986) theory of governmentality. For Foucault the modern state and the modern autonomous individual had become intertwined in their making. The state operated in and through the individual in such a way that the dividing line between them could be not drawn without severely limiting our understanding of the scope of politics. In this sense urban politics can be viewed as coursing through the urban citizen, present in multiple forms through all her/his interactions.

We find a great deal of utility in this reading of urban politics. The false distinctions between state and society can serve to de-politicize and mystify important social injustices, such as when drug abusers of economic means can address their addictions via treatment in remote, private spa-like settings, whereas those of few economic means may be targeted as undesirable neighbours by abutters of state-sponsored group residential treatment facilities in dense urban neighbourhoods. However, does this mean that politics saturates all of urban life? When Foucault (1995) develops his reading of the panopticon prison, he argues that the design's key effect is 'to induce in the inmate a state of conscious and permanent visibility that assures the automatic functioning of power' (201). As the panopticon has been developed as a metaphorical device across the social sciences, the idea that power functions through omnipresent state/society mechanisms has become influential. Indeed we can see its influence in the relational understandings of state and power discussed above (Allen, 2004).

But is the notion that governmentality, and by extension politics, is infused into most aspects of contemporary urban life a productive perspective? This depends on what we want to designate as politics. For help in answering this question, many contributors in this collection turn to the work of French philosopher Jacques Rancière (1999, 2004, 2007). For Rancière the concept of 'politics' needs to be separated from that of 'policing'. Policing is theorized as something akin to Foucault's governmentality. It is that hegemonic set of social arrangements that serve to assign and maintain roles. It is a social law that is 'thus first an order of bodies that defines the allocation of way of doing, ways of being, and ways of saying, and sees that those bodies are assigned by name to a particular place and task' (Rancière, 1999: 29). Rancière uses the phrase 'distribution of the sensible' to capture the regulatory function of policing. The 'distribution' therefore presents 'the system of self-evident facts' (2004: 12) that has institutions perform their (expected) role, citizens behave in certain kinds of ways and the authorities act as adjudicator and repressor. Given that cities are full of contestations and struggles over distributions, responsibilities and tasks, Rancière then offers a provocative thesis since these all become processes of policing.

Rancière (1999) theorizes politics as that which policing is not. Politics is the very transformation of the police order, generated by a disagreement that 'shifts a body from the place assigned to it or changes a place's destination' (ibid.: 30). This theory of politics means that it occurs rarely. Politics are not traced upon the topographies of power and/or state. Indeed, it is the antinomy of that power. Rather, politics occurs when one group of people reject their roles within the policed social order and, in doing so, they necessarily reallocate roles and make a new police order.

Politics becomes democratic when such changes are premised on equality. Put differently, democracy can only legitimate itself via equality. Political claims are therefore concerned with a party in society recognizing itself as an unequal participant: '... politics exists wherever the count of parts and parties of society is disturbed by the inscription of a part of those who have no part' (ibid.: 123). Democratic politics are therefore disruptive:

> Politics exists because those who have no right to be counted as speaking beings make themselves of some account, setting up a community by the fact of placing in common a wrong that is nothing more than this very confrontation, the contradiction of two worlds in a single world. (ibid.: 37)

An implication of this theory of democratic politics is that we cannot equate politics with either state institutions or everyday contestations:

> Democracy is not the parliamentary system or the legitimate State. It is not a state of the social either, the reign of individualism or of the masses ... Democracy is more precisely the name of a singular disruption of this order of bodies as a community. (ibid.: 99)

Democratic societies are therefore those that allow the potential for politics; the potential for the signification and resolution of inequality. A society that presumes that all are equally included is a post-democracy society. If we relate this theory of politics to urban politics, we find not that politics streams through the city. Rather, we find that politics hinges on the production of dissensus (i.e. a rejection of roles) and transformations of social order. Politics is therefore a particular type of struggle and contestation.

These various understandings of politics again leave us with the need to consider their (in)compatibility. Throughout the collection you will find various interpretations of politics. Some authors make these explicit, while other theories are implicit. We find the contrast between the 'politics are everywhere' and Rancière's politics are irregular a productive entry point for considering the politics in urban *politics*. Rancière elevates politics to a particular place within democratic societies. Politics emerges as a societal commitment to correct a wrong (i.e. an inequality within a community of equals). To label all contestations as political is therefore to lose a focus of how the particularity of politics in democratic societies. Of course Rancière recognizes there are many forms of struggle and contestation that are crucial within cities. But they might not all be political. Many are routine aspects of what he terms the police.

Here, then, we get a glimpse of Žižek's parallax shift. The ideas that 'politics are everywhere' and politics is irregular moments of dissensus are incompatible. Hence the fact that Rancière goes to great lengths to distinguish the two. His explanation provides an alternative to a parallax: it acknowledges the pervasiveness of a form of political (that he calls *la politique*) but he insists it produces orderings, not politics. So we can ask what the shift between perspectives means. For those who adopt the 'politics are everywhere' frame, we might see a multitude of worthy contestations across the city that might collectively amount to significant societal change, or individually create change in valuable ways. For those using the Rancière framing many of these contestations might be considered as events within the police order. The term politics is therefore reserved for a particular type of contestation and social change – a disruption of the police and its replacement – that maintains the prospective of a specific wrong (i.e. an unequal participant) changing society. By not subsuming politics within policing, the question of whether certain contestations deserve to become political (i.e. is it about an inequality) remains.

Structure of the Book

In light of the thinking on the urban and politics above, this collection of essays offers a selection of approaches to identify and understand urban politics. It has not been our intention to provide a text from a particular viewpoint or disciplinary perspective. Rather our attempt is to capture urban politics in its numerous dimensions. This is not to say we are disinterested in a conversation about the correct way to think the urban or politics. Indeed part of our desire to produce a text that shows urban politics in its diversity is motivated by the want for a conversation about the most just and productive ways to think urban politics. As the reader moves through the chapters and encounters different *critical* approaches, you might ask yourself what seems the most compelling in certain contributions. Do those authors who maintain a bounded conceptualization of the city better capture important social struggles? Or do those that have a very specific idea of politics provide a more insightful way to get at the social problems and their related solutions?

The chapters are organized into three sections. Each section is thematically organized according to a different way in which one can think about the city being a political space: (a) setting, (b) medium and (c) communities. We will explain these themes at the start of each section of the book. But at this point it should be stressed that these conceptualizations are not seen as distinct and discrete. Rather, they represent different theoretical perspectives (Žižek, 2006) that are able to capture the different socio-spatial dimensions of the urban.

In some cases, such as 'setting', there are clearly scalar and networked geographies that are intricately connected to the constitution of the physicality of the city. For example, the city can be a setting for municipal politics (e.g. collective provisioning of services) and a setting for capital investment (e.g. spatially targeted investment and lending). The city is therefore a setting for the enactment of various practices, whether they are scalar, such as neighbourhood (i.e. neighbourhood watch group) or metropolitan based (e.g. city government), or networked (e.g. geographies of mortgage lending). Here, the city is simply designated as a space for some practice to proceed. The city as medium recognizes that the urban is seen as a domain in which politics is identified and designated as occurring through (Foucault, 1995). This conceptualization is drawn from work in geography that has recognized how urban development reforms such as electronic surveillance (Graham, 1998) and public space design (Howell, 1993) have been motivated by particular transformations in social and political thought. Finally, the city is seen as a place of community, both in terms of a place where communities are constituted and where the politics of community are recognized and played out.

References

Allen, J. (2004) 'The whereabouts of power: politics, government and space', *Geografiska Annaler: Series B, Human Geography*, 86(1): 19–32.

Allen, J. and Cochrane, A. (2010) 'Assemblages of state power: topological shifts in the organization of government and politics', *Antipode*, 42(5): 1071–1089.

Brenner, N. (2001) 'The limits to scale? Methodological reflections on scalar structuration', *Progress in Human Geography*, 25(4): 591–614.

Brenner, N. (2004) *New State Spaces: Urban Governance and the Rescaling of Statehood*. Oxford University Press: New York.

Castells, Manuel (1977) *The Urban Question: A Marxist Approach*. MIT Press: Cambridge, MA.

Davies, Jonathan and Imbroscio, David (eds) (2009) *Theories of Urban Politics*, Second Edition. Sage: London.

Featherstone, David (2008) *Resistance, Space and Political Identities: The Making of Counter-Global Networks*. Wiley-Blackwell: Oxford.

Foucault, Michel (1986) *Disciplinary Power and Subjection*. New York University Press: New York.

Foucault, Michel (1995) *Discipline and Punish: The Birth of the Prison*. Vintage: New York.

Graham, S. (1998) 'Spaces of surveillant-simulation: new technologies, digital representations and material geographies', *Environment and Planning D: Society and Space*, 16(4): 483–504.

Harvey, D. (1987) 'Flexible accumulation through urbanization: reflections on 'post-modernism' in the American city', *Antipode*, 19(3): 260–286.

Harvey, D. (1989) 'From managerialism to entrepreneurialism: the transformation in urban governance in late capitalism', *Geografiska Annaler B*, 71(1): 3–18.

Harvey, D. (1996) *Justice, Nature and the Geography of Difference*. Blackwell: Oxford.

Hooghe, L. and Marks, G. (2003) 'Unraveling the central state, but how? Types of multi-level governance', *American Political Science Review*, 97(2): 233–243.

Howell, P. (1993) 'Public space and the public sphere: political theory and the historical geography of modernity', *Environment and Planning D: Society and Space*, 11(3): 303–322.

Hunter, Floyd (1953) *Community Power Structure: A Study of Decision Makers*. University of North Carolina Press: Chapel Hill, NC.

John, P. (2009) 'Why Study *Urban* Politics?' in Davies, J. and Imbroscio, D. (eds) *Theories of Urban Politics*, Second Edition. Sage: London. pp. 15–24.

Kjaer, A.M. (2009) 'Governance theory and the urban bureaucracy' in Davies, J. and Imbroscio, D. (eds) *Theories of Urban Politics*, Second Edition. Sage: London. pp. 137–152.

Lefèbvre, Henri (1991) *The Production of Space*. Blackwell: Oxford.

Leitner, H. (2004) 'The politics of scale and networks of spatial connectivity: transnational interurban networks and the rescaling of political governance in Europe' in Sheppard, E. and McMaster, R. (eds) *Scale and Geographic Inquiry*. Blackwell: Oxford. pp. 236–255.

Logan, John and Molotch, Harvey (1987) *Urban Fortunes: The Political Economy of Place*. University of California Press: Berkeley, CA.

Marcuse, Peter and van Kempen, Ronald (eds) (2000) *Globalizing Cities: A New Spatial Order*. Blackwell: Oxford.

Marston, S., Jones, J.P. and Woodward, K. (2005) 'Human geography without scale', *Transactions of the Institute of British Geographers*, 30(4): 416–432.

Massey, D. (1991) 'A global sense of place', *Marxism Today*, 38: 24–29.

Massey, Doreen (2005) *For Space*. Sage: London.

Massey, Doreen (2007) *World City*. Polity Press: Cambridge.

McCann, Eugene and Ward, Kevin (eds) (2011) *Mobile Urbanism: Cities and Policymaking in the Global Age*. University of Minnesota Press: Minneapolis, MN.

Nash, Gary (1979) *The Urban Crucible: Social Change, Political Consciousness, and the Origins of the American Revolution*. Harvard University Press: Cambridge, MA.

Oliver, J. (2000) 'City size and civic involvement in metropolitan America', *American Political Science Review*, 94(2): 361–373.

Pierce, J., Martin, D. and Murphy, J. (2011) 'Relational place-making: the networked politics of place', *Transactions of the Institute of British Geographers*, 36(1): 54–70.

Rancière, Jacques (1999) *Disagreement: Politics and Philosophy*. University of Minnesota Press: Minneapolis, MN.

Rancière, Jacques (2004) *The Politics of Aesthetics: The Distribution of the Sensible*. Continuum: New York.

Rancière, Jacques (2007) *The Hatred of Democracy*. Verso: London.

Saunders, Peter (1981) *Social Theory and the Urban Question*. Routledge: London.

Stone, Clarence (1989) *Regime Politics: Governing Atlanta, 1946–1988*. University Press of Kansas: Lawrence, KS.

Swyngedouw, E. (1997) 'Neither global nor local: "glocalization" and the politics of scale' in Cox, K. (ed.) *Spaces of Globalization: Reasserting the Power of the Local*. Guilford Press: New York. pp. 137–166.

Taylor, P. (1982) 'A materialist framework for political geography', *Transactions of the Institute of British Geographers*, NS7: 15–34.

Taylor, P. (2004) *World City Network: A Global Urban Analysis*. Routledge: London.

Ward, K. and McCann, E. (2011) *Mobile Urbanism: Cities and Policymaking in the Global Age*. University of Minnesota Press: Minneapolis, MN.

Žižek, Slavoj (2006) *The Parallax View*. MIT Press: Cambridge, MA.

SECTION 1
CITY AS SETTING

INTRODUCTION TO CITY AS SETTING

Mark Davidson and Deborah Martin

Setting [set-ting] ~ the surroundings or environment of anything

The chapters in this section all speak to the idea of the city as a setting for urban politics. That is to say, the city is thought of here as somewhere that politics occurs. We can, of course, say that this is a truism of urban politics, that a city is *where* we find urban politics. Our point within this section is to explore the ways in which places appear within urban politics: how the city, or particular sites within cities, function as settings for politics. We offer three takes on this question. First, we look at the idea that urban politics takes place within neighbourhoods, imagining the city as a conglomeration of different placed-based communities that collectively constitute the city. Second, we look at the city through the lens of it being a metropolitan political community. Here we are on familiar terrain for those traditionally interested in urban politics. Finally, we look at the idea that the city is a setting within a network of relations. These all represent different entry points from which to look at the city, but as you will see they are far from incompatible viewpoints.

So why concentrate on the idea of setting? The word setting refers to the environment of something. For example, you might say that a garden is a wonderful setting for a party, or that a certain public square is a great setting for a protest. In terms of urban politics, we might therefore say that the urban environment is a great setting for particular forms of politics; that, for example, the city is suited to political organization and action based on spatialized sets of concerns (e.g. neighbourhood amenities and/or environmental qualities). Or we might say that certain types of concerns, such as transportation or housing, are best dealt with at the city level. Viewing cities as settings for urban politics anticipates certain issues as urban politics, but dismisses others as unsuitable for such a setting.

This reading of urban politics aligns with comments like the following from Manual Castells:

> Cities in our societies are the expressions of the different dimensions of life, of the variety of social processes that form the intricate web of our experience. Therefore people tend to consider cities, space, and urban functions and forms as the mainspring for their feelings. This is the basis for the urban ideology that assigns the causality of social effects to the structure of spatial forms. (1983: 326)

What Castells describes here is the idea that cities themselves produce a certain politics, through both their function and related organizational needs *and* the ideological and visceral environment. We therefore get a sense of urban politics as distinctive from other kinds of politics.[1] The neighbourhood and city environment might then have their specific concerns and features that distinguish them. A task for scholars might then be to try and understand this uniqueness and compare it to other forms of politics.

This viewpoint, that cities produce a certain type of life experience and consequently generate a particular form of politics, relates back to some of the very earliest attempts to understand the industrial city. In the writings of the early urban sociologists, we see descriptions of this life experience. For example, in his 1887 description of the impersonal competitive environment of the industrial city – what he called *Gesellschaft* – Ferdinand Tonnies explains urban society as 'essentially separated in spite of all uniting factors' (74). He goes on to draw an equivalency between human relations and commodities by explaining how, in the industrial city, the former are regulated by laws not interpersonal commitments:

> Gesellschaft, an aggregate by convention and law of nature, is to be understood as a multitude of natural and artificial individuals, the wills and spheres of whom are in many relations with and to one another, and remain nevertheless independent of one another and devoid of mutual familiar relationships. This gives us the general description of 'bourgeois society' ... (87)

From this description, as well as the works of other early sociologists like Georg Simmel, Max Weber and Emile Durkheim, we find not only the idea that cities produce particular political concerns, but also that they produce particular types of people and societies. For many urban theorists attempting to make sense of the industrial city in its infancy, it was a concern for the fragmented, anonymous and individualistic nature of this society that troubled them.

> **BOX S1   THE URBAN CITIZEN IN HISTORICAL PERSPECTIVE**
>
> At the turn of the twentieth century, the first urban scholars were attempting to make sense of the vast social transformations associated with industrialization and concomitant urbanisation. German sociologists such as Ferdinand Tonnies and Georg Simmel attempted to characterize the type of person being moulded by these new environments. Famously, Simmel described urban dwellers as having a blasé attitude within the anomie of the metropolitan environment. Indifference and self-interest were commonly associated with those people living in cities. It is therefore easy to see these early urban commentators as viewing urban politics and urban dwellers in a generally negative light. However, a close reading of these works reveals a more paradoxical viewpoint. With old communal bonds being discarded, seminal urban sociology also recognized the emancipatory potential of the city. Citizens could now form new senses of self, see new social relations and form new political associations. For more on this see the following works:
>
> Adair-Toteff, C. (1995) 'Ferdinand Tonnies: Utopian Visionary', *Sociological Theory*, 13(1), 58–65.
> Frisby, D. (1994) 'Georg Simmel: First Sociologist of Modernity', in D. Frisby (ed.) *Georg Simmel: Critical Assessments*. London: Routledge, pp. 325–349.

Yet it must be noted that troubling concerns were paired with an acknowledgement of the city's emancipatory potential. In his famous 1903 essay 'The Metropolis and Mental Life', Georg Simmel describes the freedoms brought about by the anonymity of commodity-based urban societies. Transcending familial and tribal bonds brought with it the potential for the urban dweller to construct new identities and bypass social strictures. We might say the same about the political concerns that Castells claims are generated by the city. The advent of urban infrastructures, such as sewers, roads, railways and so on, may well have brought with them political conflicts. But at the same time they generated, as least for some, liberation from disease and distance. As you read through the pages of this section, it might well be worth keeping the two sides of the coin in mind. The social conflicts and struggles generated in the city certainly have many problematic elements. But might these conflicts and struggles be generative of new opportunities and modes of life? And how might we gauge if these new opportunities are indeed generated?

The first chapter in this section written by Katherine Hankins and Deborah Martin approaches the question of city as setting by looking at the neighbourhood as a space for politics. They argue that the neighbourhood has long been understood as a powerful generator of political movements. But the idea that the neighbourhood is a contained space is rejected in favour of a relational understanding of neighbourhood. Here neighbourhood-based politics are viewed as emanating from 'flexible and necessarily multi-scalar' sets of processes. As a result, Hankins and Martin reject the idea that neighbourhoods have self-contained politics. Rather, they prefer to see neighbourhood politics as a bundle of processes that become manifest in place. In order to develop this understanding of the neighbourhood setting, they draw upon a study of a faith-based, neighbourhood-based social movement: 'strategic neighbouring'. Through their description of this missionary-type urban endeavour, Hankins and Martin show both the networked and located features of neighbourhood politics. But drawing on Rancière's understanding of politics, they claim that strategic neighbouring never achieves a true politics. Here they point towards the inability of the strategic neighbouring movement to see beyond the politics of the neighbourhood, or city for that matter. Thus, whilst Hankins and Martin acknowledge the importance of neighbourhood setting for the development of some urban politics, they argue that politics which remain fixed at the neighbourhood scale may constrain ambitions and goals.

In the following chapter Kevin Ward looks out of his window to examine the city as a political entity. This begins by reviewing the major theories that have explained 'urban politics' over recent decades. The theory that Ward describes as 'old' is that associated with Manuel Castells' idea that urban politics are those related to issues of collective consumption. This is contrasted to 'new urban politics' (Cox, 1993). Rather than being concerned with collective provisions (e.g. roads, schools, trash collecting) this urban politics is based squarely upon economic development. In a world of economic competition, cities have transformed the ways they are governed in order that they achieve requisite economic growth. But, as Ward explains, this is not to say we can consider the city as discrete entity that competes with other cities. Rather we might need to see the city as 'splintered' in that different policies and strategies shift from city to city. Using the example of Calgary, Ward narrates how the tentacles of American think-tanks stretched into, and shaped, the city's planning process. The chapter leaves us with the need to think 'urban politics' in both scalar and relational senses. Scalar in the sense that the city legislates for itself with in boundaries (i.e. Calgary city limits) and relationally since this legislation emerges from a whole network of connections to policy making around the globe. But do we have an adequate theory of urban politics to capture both these geographical understandings of the city?

The final chapter in this section by Elvin Wyly and Kathe Newman explores the idea of urban politics as networked through a rich empirical account of mortgage foreclosures in the US. Since 2007 waves of housing foreclosures have swept across the US, causing a financial house of cards to collapse across the globe. As Wyly and Newman explain, these foreclosures have had a distinctly localized geography, being clustered in certain types of, often poverty-stricken, neighbourhoods. We might then view the foreclosure phenomenon on as a localized one with global implications. But what Wyly and Newman demonstrate is that foreclosures have to be viewed as bound up with national and international financial networks. Predatory lenders had been targeting particular neighbourhoods in order that the goals of national and global financial actors were served. Here, then, the city is a setting for urban politics in the sense that localized processes are now always bound up in a global political economy, the story of the foreclosure crisis being not one of poor people borrowing irresponsibly and rogue lenders, but rather a significant co-constitution of capital and community.

Through the different entry points of 'neighbourhood', 'city' and 'network' the chapters in this section therefore challenge us to think carefully about the geographical lenses we use to understand 'urban politics'. What is shared across the three chapters is an attempt in each case to incorporate different geographical lenses: neighbourhood + national organization, city + global policy networks and local community + global capital accumulation. Whilst all maintain the idea of the city as a setting of urban politics – the city as an environment where particular politics play out – they also place this 'container' of urban politics within a relational setting. The consequences for political thought and action are multiple. For Hankins and Martin it means we must understand what goes on within a neighbourhood as a consequence of the multiple intersecting relations that constitute it. Consequently, they urge us to rethink about what constitutes political action in any neighbourhood. For Ward it means that the politics of right-wing think-tanks must be implicated in debates over how to grow a city. And for Wyly and Newman it means we cannot understand the geography of mortgage foreclosures without understanding how it is that they are bound up with a capitalist class's attempt to accumulate wealth through the dispossession of others. None of the chapters in this section therefore rely solely on a scalar bracketing of urban politics. Rather they choose to frame urban politics as something that occurs within a particular setting, but which gets constituted through a host of processes with varied geographical qualities.

QUESTIONS TO FURTHER EXPLORE

- If all spaces are relational, do we need to bracket particular 'settings' for (urban) politics?
- How important are neighbourhood and/or city associations to you?

- What are the political consequences of seeing places like neighbourhoods and cities as relationally constituted? Who then becomes a member of a particular city or neighbourhood?
- Do we still have a particular brand of politics associated with cities, or are all politics urban?
- How do different settings – neighbourhoods, cities, network nodes – operate together?

Note

1 For an appraisal of the contemporary relevance of Manuel Castells' theories of urban politics see Ward and McCann (2006); further discussions are contained within the same issue.

References

Castells, Manuel (1983) *The City and the Grass Roots: A Cross-cultural Theory of Urban Social Movements*. University of California Press: Berkeley, CA.

Cox, K. (1993) 'The local and the global in the new urban politics: a critical view', *Environment and Planning D: Society and Space*, 11(4): 433–448.

Simmel, G. (1995) [1903] 'The metropolis and mental life', in Kasinitz, P. (ed.) *Metropolis: Center and Symbol for Our Times*. New York University Press: New York. pp. 30–45.

Tonnies, Ferninand (1955) [1887] *Community and Society (Gemeinschaft und Gesellschaft)*. Dover Publications: London.

Ward, K. and McCann, E. (2006) '"The new path to a new city"? Introduction to a debate on urban politics, social movements and the legacies of Manuel Castells' *The City and the Grassroots*', *International Journal of Urban and Regional Research*, 30(1): 189–193.

2

THE NEIGHBOURHOOD AS A PLACE FOR (URBAN) POLITICS[1]

Katherine Hankins and Deborah Martin

Introduction

Urban politics often implies a local politics, or neighbourhood politics. In this chapter, we examine the notion of politics as local or neighbourhood oriented. We explore common conceptualizations of 'neighbourhood' and argue, ultimately, that neighbourhoods should be considered as every day, momentary and instanciated bundles of place identities and relationalities (Pierce et al., 2011) which requires us to question whether neighbourhood politics are necessarily local. We think through neighbourhood as a *relational place* that has the inherent potential to invoke politics. We then problematize this potential by examining the meaning of politics, drawing from the conceptualizations of 'the police' and 'politics' as expressed by Jacques Rancière. Drawing on our case study of 'strategic neighbours', in which people of faith settle in low-income communities as a way to serve them, we conclude with a call to destabilize neighbourhood in favour of place politics that seek dissensus, or a voice for those who are not heard.

Neighbourhood as place

Scholars in geography and urban studies more generally have long struggled with defining the concept of *neighbourhood* (Park et al., 1967[1925]; Hunter, 1979; Olson, 1982; Galster, 1986, 2001; Kearns and Parkinson, 2001; Clark 2009). Far from having given up on this vague concept, a special issue of *Urban Studies* (2001) demonstrates continuing interest among scholars in defining, applying, and investigating the significance of 'neighbourhood' for individual behaviour, health, and life chances (Buck, 2001; Ellaway et al., 2001; Galster, 2001); in fostering social allegiances, identities, and capital (Forrest and Kearns, 2001; Kearns and Parkinson, 2001; Purdue 2001); as an indicator of urban growth and change (Butler and Robson, 2001; Galster, 2001); and, importantly for our discussion, in shaping political decisions and structures (Allen and Cars, 2001; Docherty et al., 2001). Neighbourhoods also have an increasingly political meaning and function in the neo-liberal era, in which governments seek solutions to social and economic problems by devolving responsibility – without resources – to more local areas (Raco, 2000; McCann, 2001; Meegan and Mitchell, 2001; Elwood and Leitner, 2003; Newman and Ashton, 2004; Purcell 2008).

Martin (2003a) draws upon an extensive literature review to argue that we do not know neighbourhoods when we see them; we construct them, for purposes of our research or social lives, based on common ideals of what we expect an urban neighbourhood to be. The neighbourhoods that we define through research or social exchange are always subject to redefinition and contention; they are not self-evident. A neighbourhood is a type of place, and, as such, should be studied as a contingent, flexible space that nonetheless has material, experiential salience for people's lives. Neighbourhoods may be like any other type of territorially based social ideal, in that they are socially as well as spatially constituted, and are, as Anderson (1991) suggested in reference to nations, 'imagined' by those who share them (e.g. Cope, 2008).

Given definitions of neighbourhoods as sites of daily life and social interaction (e.g. Hunter, 1979; Galster 1986, 2001; Forrest and Kearns, 2001), we suggest that neighbourhoods are a particular type of place: locations where human activity is centred upon social reproduction (see Castells, 1977, 1983); or daily household activities, social interaction, and engagement with political and economic structures. Neighbourhoods derive their meaning or salience from individual and group values and attachments, which develop through daily life habits and interactions. Neighbourhoods, like places, are 'where everyday life is *situated*' (Merrifield 1993a: 522; *emphasis in original*). Furthermore, as Schmidt (2008) suggests, neighbourhoods can gain their meaning through sustained practices that produce them in particular ways over time.

If neighbourhoods are places, we can examine them as a particular form of that geographic concept. Agnew's (1987, 1989) definition of place as locale (site of daily life), location (a site with connections and relations to broader social, political, and economic processes at varying scales), and sense of place (affective feelings) captures the many facets of neighbourhood that other scholars have identified (Park et al., 1967; Hunter, 1979; Galster, 1986, 2001; Forrest and Kearns, 2001). Escobar (2001), writing about place, argued that places are constituted through two processes: political economy and humanistic sense of place. Political economy shapes places through local and global economic processes of capital investment, while sense of place reflects the sentiments people feel about a place, derived from individual experiences, attachments, and social connections. These two processes roughly parallel Agnew's location and sense of place categories, but only implicitly includes locale, as the meeting-point of location and sense of place. Nonetheless, both views capture the combination of economic processes and individual, cognitive attachments in shaping place. These are fundamental elements of places, but need to also be considered as always flexible and simultaneously multi-scalar, rather than necessarily local (Pierce et al., 2011).

Hunter (1979) characterized neighbourhood as 'a uniquely linked unit of social/spatial organization between the forces and institutions of the larger society and the localized routines of individuals in their everyday lives'. For Hunter, the context of the neighbourhood – its linkages with other places, or within places – ought to be part of any analysis of neighbourhood. This recognition of the embeddedness, and, therefore, of the multiscalar nature of neighbourhoods within a larger set of routines and social, political, and economic forces, is one that echoes the approach of Suttles (1972). He argued that neighbourhood can mean the immediate home area, the locality of a few blocks, and/or the entire urban region (Suttles, 1972, cited in Kearns and Parkinson, 2001). Conceptually, we draw from Pierce et al. (2011) to suggest that neighbourhoods are *relational* places. Rather than conceiving places or specifically neighbourhoods as location, locale, and sense of place, relational place embeds social and political-economic relations with affective and environmental features as 'bundles' (drawing on Massey, 2005). These bundles (places) develop simultaneously agentically and structurally, as they are both individually experienced and socially expressed and lived (Pierce et al., 2011). Thinking about neighbourhoods as relational places requires serious consideration of the ways in which they are embedded in and connect to urban politics.

Neighbourhood and Politics

The utility of the concept of 'neighbourhood' for much contemporary urban political geography derives from its construction through political strategy and contestation. The ideal of neighbourhood asserts a role for the 'local' in a world increasingly characterized by extra–local interactions and exchanges. The locally based activism that can occur in neighbourhoods, regardless of the particular motivation or cause, demonstrates the important role of local areas in situating the grievances that form a basis for activism (McCarthy and Zald, 1973, 1977; Traugott, 1978; Escobar 2001). For example, community organizations in Chicago use geographic information system (GIS) analyses and mapping in order to position their neighbourhoods as sites of resources and need in local municipal policy discussions (Elwood, 2006). In Los Angeles, residents of the San Fernando Valley worked to preserve a neighbourhood landscape of single family homes through political activism aimed at secession from the broader city, prioritizing a 'local' suburban identity against the broader city (Purcell, 2008).

Political agendas and concerns can coalesce in the particular spatial locations of the actors involved in conflicts. Thus, neighbourhoods are often formed and constituted through activism, sometimes in response to imposed boundaries or efforts to delineate new boundaries. Robinson (2001) investigated the reaction of residents to a proposed road in Glasgow that would have severed the connection of one residential district to a local park, further spatially constraining an already economically disadvantaged community. Residents of the area expressed 'fears of exclusion and segregation' about the proposed land-use change (Robinson, 2001: 101). In that case, a land-use change resulted in a new, more rigid boundary for a neighbourhood. Likewise, activism around the construction (or demolition) of institutions such as schools, parks, or public housing further develop neighbourhood boundaries. For example, parents in a gentrifying area of Atlanta, Georgia, created a charter school, which required contending with neighbourhood identities to draw attendance zone boundaries, and in the process solidified the territorial extent of the neighbourhood (Hankins, 2007). Neighbourhood change, such as gentrification processes or selective redevelopment in poor neighbourhoods, can foster the emergence of class-based forms of neighbourhood politics, creating intra-neighbourhood tensions (Newman and Ashton, 2004; Hankins, 2007; Martin, 2007).

The scale at which neighbourhoods are defined can also be the basis of dispute, however, where residents of an area may define their spatial community at a different scale than the perspective of local public officials. McCann (2003) shows how city-wide concerns over sprawl and growth in Austin, Texas, were translated into new neighbourhood-based planning and zoning programmes. The city sought to increase demand for and densities of housing in the urban core by fostering more intensive land uses and revitalizing the landscape. Some residents of the affected neighbourhoods resisted the small-area, neighbourhood-based focus of planning efforts in favour of larger, regional coalitions of poor and mostly Latino neighbourhoods in order to fight what they perceived as White gentrification into their neighbourhoods. These residents defined their communities in terms of economic, ethnic, and locational criteria, and their definition of 'neighbourhood' was at a broader scale than that of the city planners.

A relational approach to neighbourhood and urban politics anticipates the flexibility of identities and territorial affiliations that McCann's case highlights: people affiliate and recognize themselves as members of communities that are linked to particular, material, and tangible locations but which are also simultaneously connected spatially, socially, politically, and/or economically. As such, we suggest that neighbourhood-based politics can be more constructively thought of as *place* politics.

Geographers have spent part of the past few decades refining our understanding the relationship between politics and place, and more specifically the political possibilities of conceptualizing sociospatial dimensions of place (Martin, 2003a; 2003b; Massey, 2004, 2005; Jessop et al., 2008; Leitner et al., 2008; Pierce et al., 2011). The various contours of place have been examined, turned over, left for dead in some cases (alongside 'militant particularisms' (Harvey, 1996)), and resurrected (Massey, 1991, 1994, 2005). Massey (2005: 119) suggests place is 'the collection of interwoven stories', and 'a bundle of trajectories', which represent the stories and experiences of people who interact in and with particular space-times. Furthermore, McCann and Ward (2011) suggest that places are 'assemblages of elsewhere' which create and mobilize urbanisms through global policy transfers. They focus on the territorializations of urban policy in particular places through global exchanges, situating urban politics itself as a global, though locally instanciated, phenomenon.

The place/politics juncture has often been empirically focused on the moment of negotiation and/or contestation over places or place identity. According to Pierce et al. (2011), politics are the processes of negotiation over the terms that govern the use of space and place, which may include contestation over discursive place representations, scalar conceptualization or the terms of participation in space/place (e.g. Martin et al., 2003; Purcell 2008). Place/politics, then, poses the conjuncture not simply of open conflict over space or land or people, but over being in situ: 'Places pose in particular form the question of our living together. And this question [...] is the central question of the political' (Massey, 2005: 151). Politics makes bare 'the moment of antagonism where the undecidable nature of the alternatives and their resolution through power relations becomes fully visible' (Laclau, 1990: 35, cited in Massey, 2005: 151). This definition highlights the openness of possibilities of place/politics, and points to dilemmas between ordering places to resolve antagonisms, and making those antagonisms fully visible, a conflict confronted in political philosopher Jacques Rancière's understanding of politics.

One of Rancière's central concerns has been the interrogation and rethinking of democracy. This is not democracy in terms of liberal (capitalist) democracy, but rather democracy as a social order founded on a notion of egalitarianism. For Rancière (true) democracies are societies that are continually reworked by the recognition of inequalities and consequent granting of equalities that transform the said society. Rancière's (2001) conceptualization of the political within his understanding of democracy hinges on a distinction he makes between 'the police' and 'politics'. For Rancière, the police is 'an order of bodies that defines the allocation of ways of doing, ways of being, and ways of saying, and sees that those bodies are assigned by name to a particular place and task; it is an order of the visible and the sayable that sees that a particular activity is visible and another is not, that this speech is understood as discourse and another as noise' (1999: 29). For Rancière, policing partitions the social into knowable parts with attendant places. Most conceptualizations of politics or government are, for Rancière, more concerned with questions of the 'police' and its ordering of social bodies. By contrast, 'politics' is the 'intervention upon the visible and the sayable' (Rancière, 2001: paragraph 21). The essence of politics is 'to disturb this [police] arrangement by supplementing it with a part of the no-part identified with the community as a whole' (Rancière, 2001: 21). The function of politics within a democratic society is 'the configuration of its proper space. It is to disclose the world of its subjects and its operations. The essence of politics is the manifestation of dissensus, as the presence of two worlds in one' (2001: thesis 8). The two worlds are the ordered, sensible, named and the world marginalized by the naming and ordering.

Politics therefore become concerned with what, within the existing police order, is unknowable (unsayable, invisible) becoming known and sensible. When an unequal group – what Rancière calls 'the part with no part' – declares itself unequal within a system of declared equality, it therefore constitutes a radical disagreement that challenges currently assigned roles. Politics thus challenges all that is ordered in the world of the police. Politics is, therefore, wrenching, confronting, and challenging in ways that must disrupt and undermine presumed orders.

Rancière's 'police' includes, for urban politics, the ordering and naming of territories such as neighbourhoods as specific, bounded, often very local sites of participation in urban governance. Neighbourhoods define and create manageable, fixed spaces or territories for urban politics to take place. Many cities have codified the importance of neighbourhoods through formal planning processes, creating territorially designated planning units in which the 'local' scale becomes reified, as the logical scope for planning or policy solutions. Through processes designated to include residential spaces and populations in policy input, the neighbourhood becomes one of the appropriate structures/scales/spaces through which (urban) society is governed. Neighbourhood, then, becomes a site of ordering, stabilizing and partitioning the sensible. Whilst not inherently problematic given the fact that every society requires some ordering of roles, this designation of neighbourhood politics can certainly generate an inability to speak to important social issues. To put it in Rancière's terms, the ordering of neighbourhood politics might well serve as a mechanism to avoid the political; to depoliticize. If indeed this ordering does depoliticize, we would therefore need to see neighbourhood in a way that enables a politics that can potentially destabilize and transcend those orders/names that limit the neighbourhood becoming an effective locale for social change.

For Rancière place is not merely political because of the 'question of our living together' (Massey, 2005), but as Mustafa Dikeç (2005: 172) suggests, place becomes political in that it becomes the 'site where a wrong can be addressed and equality can be demonstrated. It becomes an integral element of the interruption of the natural (or, better yet, naturalized) order of domination through the constitution of a place of encounter by those that have no part in that order'. Places can become political when the noises of discontent within them become political voices that construct dissensus. This conceptualization opens up the relationship between place and Rancièreian politics. Yet this relationship is not guaranteed; place must *enable* dissensus, an articulation of a wrong. To do that, it must enable the interruption of the ordering. In what follows, we ask whether neighbourhood-based politics can be Rancièreian politics, through destabilization of the categories (of territory and residents) that neighbourhoods produce.

We draw from survey data and interviews with 'strategic neighbours' in US cities to explore the complexity of neighbouring as practice and neighbourhood as a setting for police and, potentially, Rancièreian politics. We examine the place-making and political encounters of strategic neighbours, who are middle-class people of faith relocating into some of the country's poorest urban areas. The case of strategic neighbouring highlights the conflicts between neighbourhood as a site for politics and for police, in Rancière's terms, while illustrating the relationality of the place-making that produces and is produced through urban politics.

Strategic neighbouring

'Strategic neighbouring' is a term developed by Robert Lupton (1997), one of the founders of the Christian Community Development Association (see ccda. org) that represents a broader (social) movement of faith-motivated people who serve the poor by living among them.[2] Strategic neighbours are people who move into what they term under-resourced areas, be they inner-city neighbourhoods, suburban apartment complexes, or trailer parks, to offer their social connections, emotional support, political organizing, money, and faith. This movement grew out of what John Perkins (1995), a civil rights activist, refers to as the three Rs: relocation, redistribution, and reconciliation. Relocation involves the physical move into poor areas; redistribution suggests the reallocation of resources as part of this process; and reconciliation refers to addressing racial barriers and racism. In essence, part of their mission is to confront the injustices of poverty and racism by living among and developing relationships with their neighbours.

> **BOX 2.1 STRATEGIC NEIGHBOUR CRISSY BROOKS DESCRIBES GOALS**
>
> Crissy Brooks, from Costa Mesa, California, is a 35-year-old White woman who has lived for seven years in a neighbourhood that is 98 per cent Latino. She reflects on her goals as a strategic neighbour:
>
> > I want my neighbors to be able to afford their own place to live. I want the police to treat our teens with dignity. I want my neighbors to be able to work legally. I want landlords to fix structural problems. I want people outside my neighbourhood to know my neighbors.
>
> For Crissy, her issues involve immigration, which reflects the different kinds of issues across different place contexts. She also explicitly seeks to resignify the meaning of her neighbourhood to those beyond its borders.

> **BOX 2.2 STRATEGIC NEIGHBOUR SCOTT DEWEY DESCRIBES DIALOGUE AS A GOAL OF STRATEGIC NEIGHBOURING**
>
> Scott Dewey, a 48-year-old White man who lives in the Whittier neighbourhood of Denver, Colorado, which is 30 per cent African American, 40 per cent Latino, 30 per cent White, suggests that his goals are 'Understanding and dialogue between new gentry (usually White), new immigrants (usually Latino), and long-time residents (usually African American)'. Scott has lived in five different neighbourhoods over the past 25 years – and his location in Whittier seeks to mitigate the deleterious effects of gentrification.

> **BOX 2.3 STRATEGIC NEIGHBOUR ASHLEE STARR DESCRIBES HER ATLANTA NEIGHBOURHOOD**
>
> Ashlee Starr, a 27-year-old, White, married mother of two children has lived for five years as a strategic neighbour, and most recently for the past year in the

neighbourhood of Pittsburgh, Atlanta, which is 98 per cent African American, where over 50 per cent of homes are vacant. She expresses her understanding of the conditions of her neighbourhood space and what she, as a strategic neighbour, has sought to do about it:

> The physical conditions are awful because no one is being held accountable to deal with the mess (banks, investors, etc.). We only have one park in the neighborhood, so kids are left to play in the street which effects *[sic]* their environmental conditions. We have a large number of homeless men and women living in the neighborhood who need social service help. Prostitution is a MAJOR problem throughout the entire neighbourhood as well.
>
> We are here to be a safe place and to love those around us. We want to be a part of this neighbourhood. The same struggles that our neighbors are facing, we are also facing. We aren't involved on the macro level but the micro level.
>
> We allow the homeless to be comfortable in our home, we let kids do homework on a shared computer in our dining room, our swing set is open to all. We don't have a lot of answers and we pray daily for others to want to move in to our very abandoned neighbourhood and feel the same call we do. We want to be here and be present. We are still working on what that looks like.

Ashlee reflects on the struggles of strategic neighbouring as developing relationships with those who are marginalized or left behind in her largely abandoned neighbourhood. She, like other strategic neighbours, embody place and place-making and in the process, their subjectivities are constructed by the negotiation between 'police' and politics.

We want to suggest that, in fact, these strategic neighbours intentionally reconfigure the bundles of the neighbourhoods in which they live, connecting their networks to the networks of incumbent neighbours (see Hankins and Walter, 2012). In the process, many strategic neighbours have grappled with their own social positioning and have engaged in various channels of the police, formal neighbourhood associations or city hall, for example, and, based on our interviews, have suggested a deep puzzlement at the ineffectiveness of these formal channels of politics – what we interpret as a recognition of the ways in which these structures of policing do not address the needs of the poor and socially marginalized. The ineffectiveness that strategic neighbours encounter is, to put it again in Rancière's terms, the experience of the partitioning of the sensible that leaves a part of no part. For strategic neighbours, the result of such a confrontation is that they see themselves as caught between the officially sanctioned political process of the neighbourhood, such as neighbourhood organizations or city hall, and the realities of deep poverty and marginalization that keep many of the poorest and disenfranchised neighbours from participating in these channels. In Rancièreian terms the act of relocation creates fissures in what is sensible for strategic neighbours; they see the gaps and inconsistencies, the inequalities of the current police order.

Our case study draws on an 18-month study of both urban ministries and strategic neighbours in North America, and specifically in the Atlanta, Georgia, area.[3] We conducted an online survey with approximately 70 strategic neighbours from across the United States from 18 states and from Canada: 36 women and 34 men; 23 per cent were single with a median age of 29.5 and a mean age of 35; 84 per cent self-identified as White, 7 per cent as non-White. Furthermore, we conducted three focus groups in Chicago, Illinois, and in Atlanta, Georgia, and carried out interviews with 37 strategic neighbours who live in some of Atlanta's poorest neighbourhoods.

Strategic neighbours seek to redirect the place trajectories (Massey, 2005) of the neighbourhoods in which they live by connecting their (multi-scaled) resources to their neighbours, in the form of friendship, social support, financial contributions, and engagement with formal governance structures. Barbara Fiske, a 40-year-old Latino stay-at-home mother, who has been a strategic neighbour for 20 years in Fresno, California, reflects on how the neighbourhood's needs should be met:

> Ah. With loving neighbors. More strategic relationships. Relationships are what is going to cause change. Not so much programs or events. Our family's personal vision is to be available when 'life' happens. We build friendships and trust. Currently, when we engage with neighbors who are going through serious, intense drama/crisis in their lives we have such a network of support and resources.

She continued, commenting on the different connections she has across various 'scales' of the city:

> With neighbors, working on our front lawn gardening. Being introduced by other neighbors. At events such as block parties, school fairs, etc. [We work] with government agencies – because of the amount [*sic*] of strategic neighbours here and their longevity we have partnered with City Hall to bring many improvements and maintained weekly meetings to address current issues. [We work] with churches and faith-based organizations – there is quite the networkings and meshing of resources to address the needs in the neighbourhood.

Barbara reveals an emphasis on connecting neighbours within the neighbourhood and providing connections to city services and to broader networks of faith-based organizations and churches. She embodies an extensive network, connecting her neighbourhood to resources such as the institutions of her faith and of government, well beyond its borders. Although these institutions seek to transform the circumstances of the poor neighbourhoods that they serve, they nonetheless are part of a Rancièreian police, an ordering that, while allowing poverty to be visible, does not necessarily interrupt and contest that ordering.

Figure 2.1 Thanksgiving Day on Cheryl's front porch, Booker T. Washington neighbourhood, 11 November 2010. Cheryl has been a strategic neighbour for four years in Atlanta (source: Cheryl Case)

Many of the strategic neighbours we interviewed sought to reconfigure the bundles of their place-networks by changing the material spaces of the neighbourhood and fostering relationships among neighbours. For example, Michael Wong, a 26-year-old Asian man, who has lived for three years as a strategic neighbour in a New Orleans neighbourhood that is 98 per cent African American, is in the process of developing a house for interns in addition to creating a 'tutoring/safe space for [recreation] and work'. He is also developing 'a community garden to increase neighbour connectivity' and reflects that he and his wife 'intentionally spend time walking around outside getting to know our neighbours'.

These efforts to get to know people, really connect with them, illustrates the malleability of the neighbourhood as a site that is not solely local, but that nonetheless provides common ground, literally, for people to meet and interact. In relational terms, the neighbourhood is a site where multi-scalar processes, from sidewalk hellos to city hall and global economic relations, converge and are expressed. At the same time, they are sites of politics – both the Rancièreian police partitioning, as well as a nascent politics of disruption, of open incongruity between the sensible and the invisible.

Barbara and other strategic neighbours revealed frustrations with the power relations embedded in their social positioning and also with the ineffectiveness of many of their (middle-class) political channels to effect change (see Case, 2011). When asked directly about the degree to which strategic neighbours see their work as political (which we did not define), the answers were quite mixed. From our survey, 30 per cent of respondents identify their work as political; 70 per cent suggest it is not. When asked to explain their position, strategic neighbours offered a variety of answers. Michael Wong (from New Orleans) stated 'I don't have a political agenda, but I know that politics do have a lot to do with the needs of the neighbourhood, and that you have to be somewhat involved in politics to get things done.' For Michael, it is formal state channels, such as neighbourhood associations, city hall, or being involved in political campaigns, that he defines as politics. For Rancière, of course, these are classically the police that partition the sensible. The neighbouring that Wong does is not, for him, political, even though through his actions (of creating an intern centre, or building social connections) he seeks to change the material and social dynamics of the community. Indeed, he seeks to foreground the part that has no part, and in doing so, he pushes forwards (but does not necessarily create) dissensus, a making visible of that (who) which previously was not.

Some of our survey and interview participants initially eagerly embraced the neighbourhood association – the formal structure of police – as a vehicle for change for the neighbourhood. But then, after months and generally within the first year, they became frustrated with the class structures embedded in the neighbourhood associations. For example, Samantha Greg, White mother of two who was a strategic neighbour for two years in the Vine City neighbourhood of Atlanta, elaborated on the complexity of being involved in 'politics' (police):

> We used to go to Civic Meetings but don't anymore. I think we helped for 6 months or more. We weren't super involved, helped with some studies and things but we just waned out ...
>
> That is not to say the civic association is bad, but it was frustrating. The people we had grown to love were not respected ... It was like if you rented and didn't own then you weren't a part of the community. And 85% of the community at the time were renters ...

Figure 2.2 Community garden in English Avenue, west Atlanta, 11 October 2009. This community garden was developed by residents and strategic neighbours

The frustrations of strategic neighbours with the formal politics of neighbourhood associations and interactions with city government illustrates the embodied conflict of the police (or the ordering of state channels) and politics – or the arena where the voice of the marginalized can be heard. For neighbours like Samantha Greg, politics were not really possible through the neighbourhood association, which clearly did not offer a voice to those who were marginalized in the poverty of Vine City.

Other strategic neighbours underscored the absence of marginalized innercity residents from formal governance structures. For example, Bart Campolo, a 47-year-old, White, married father of two, in the Walnut Hills neighbourhood of Cincinnati, Ohio, plainly stated that his involvement in 'politics' is critical, as 'The folks we're with here have no money, and therefore no political voice'. His observation echoes the marginalization of the impoverished as lacking an ability to participate in what Rancière would term 'the police'. Anna Terry, 31, a White, single woman in the Binghampton neighbourhood of Memphis, Tennessee, echoed a similar sentiment but with the opposite framing: 'I'm not worried about politics. I am worried about the lives around me, sometimes that involves politics, but very rarely can politicians help me or the folks around me.' The idea that politicians cannot really help highlights Rancière's assertion that they are part of a police order that does not want to allow for politics (i.e. post-democratic); a partitioning that does not enable the powerless to have a voice or to participate in a supposed democracy.

Figure 2.3 Historic home in South Atlanta, two doors down from a strategic neighbour's residence. South Atlanta, a neighbourhood of approximately 550 homes, has over two dozen strategic neighbours (source: Katherine Hankins)

Strategic neighbours recognize how the structures of the police fail to help them in their goals of connecting with and transforming their own lives and those around them. Instead of achieving such transformation, however, frustration with the disconnection between the police and dissensus politics both constructs the complex sociospatial positionality of strategic neighbours and it also contributes to what many of them term 'burnout' (see Case, 2011). For example, Richard Humphrey, a White 30-year-old father of two, in the West End, Atlanta neighbourhood for the past eight years, contemplates his frustration:

> I saw so much change ...working in an upper middle class, white youth group setting ... But what you see here is so little change in people's lives, and what dominates most of all is seeing brokenness, seeing dysfunction, seeing kid after kid drop out of high school, seeing relationships as dysfunctional, not good for kids ... I have seen it with the kids on my basketball team. Where I have seen kids go from being a 9th grader to like a, you know, to becoming a grown man and watching 7 out of 10 of them drop out of high school, involved in some sort of crime, and then in and out of prison. And it is just kind of depressing to see that over and over again.

Almost half of the survey respondents identified 'burnout' as a serious challenge to strategic neighbouring – evaluating it 4 or 5 on a scale of 1 to 5, where 5 is the most significant challenge. This burnout around the sociospatial subjectivity of the strategic neighbour is, we suggest, constituted by the constant confrontation with police and politics – and the unsatisfactory options of how to deal with the kinds of everyday living conditions that strategic neighbours see their neighbours experience.

Conclusions

Through their insertion into inner-city life, strategic neighbours reconfigure the place bundles in the neighbourhood setting, which results in a complex sociospatial positionality (for them and the neighbours with whom they interact). As part of and through their rebundling, strategic neighbours seek to give voice to the voiceless, which in Rancièreian terms could help to produce politics. Yet they encounter the ineffectiveness of the formal governance structures situated in and through the neighbourhood. That is, strategic neighbours wrestle with contemporary forms of policing, the categorization of the parts of society into particular spaces, rather than engaging and enabling dissensus (i.e. politics). Strategic neighbours situate themselves at the juncture between police and politics: challenging and reworking police categorizations is central to their purpose, but at the same time, they recognize that the institutions seemingly designed to give voice to marginalized neighbours effectively keep the unknowable and the unsayable unknown, as suggested by Rancière (1999) in his critique of contemporary democratic societies. While everyday police partitioning recognizes the presence of poor, marginal neighbourhoods and residents, they remain to a large degree without voice (i.e. they are unequal). The limitations encountered by strategic neighbours suggest that neighbourhood itself, as an imagined political unit and meaningful territory for social change, is a limited concept for opening up politics to and for the unknown.

Neighbourhood in urban politics in this case, then, sits as part of the current post-democratic police. A relational place-making perspective on neighbourhood insists on a partitioning which is flexible and unordered; open to rebundlings which may foster a politics of dissensus, because it resists the bounding of neighbourhoods as 'local'. Strategic neighbouring – through its frustrations and confrontations with the ordering of urban politics-as-police – highlights the necessity of conceptualizing neighbourhoods as relational places, constantly in process, as a means to challenge the ordering of an urban politics definition of neighbourhood. It is through actions like the rebundling that strategic neighbours do that conscious place-making and politics may occur. Real neighbourhood politics – a politics of dissensus – require challenging fixed, bounded notions of neighbourhood and reworking them as flexible, contingent dynamic places of/for an urban politics of the as-yet-unknown.

Notes

1 Portions of this chapter appear in D. Martin (2003) 'Enacting neighborhood', *Urban Geography*, Vol. 24, No. 5, pp. 361–385. Reprinted with permission from © Taylor and Francis Group, Milton Park, Abingdon, Oxon, UK, OX14 4RN. All rights reserved. The authors gratefully acknowledge this permission from Taylor and Francis Group (www.tandfonline.com).
2 There is no official 'count' of strategic neighbours across the United States, but preliminary research suggests they are in most major cities and develop through networks such as the Christian Community Development Association, which has over 3,000 individual members.
3 This research was conducted by co-author Katherine Hankins along with Andy Walter, University of West Georgia, and Cheryl Case, Georgia State University.

References

Agnew, John (1987) *Place and Politics*. Allen & Unwin: Boston, MA.
Agnew, John (1989) 'The devaluation of place in social science', in Agnew, J.A. and Duncan, J.S. (eds) *The Power of Place*. Unwin Hyman: Boston, MA. pp. 9–29.
Allen, J. and Cars, G. (2001) 'Multiculturalism and governing neighbourhoods', *Urban Studies*, 38(12): 2195–2209.
Anderson, Benedict (1991) *Imagined Communities: Reflections on the Origins and Spread of Nationalism*, rev. edn. Verso: New York.
Buck, N. (2001) 'Identifying neighbourhood effects on social exclusion', *Urban Studies*, 38(12): 2251–2275.
Butler, T. and Robson, G. (2001) 'Social capital, gentrification and neighbourhood change in London: a comparison of three south London neighbourhoods', *Urban Studies*, 38(12): 2145–2162.
Case, C. (2011) 'Strategic neighboring and "Beloved community development" in West Atlanta neighborhoods', unpublished master's thesis, Georgia State University.
Castells, Manuel (1977) *The Urban Question: A Marxist Approach*. MIT Press: Cambridge, MA.
Castells, Manuel (1983) *The City and the Grassroots*. University of California Press: Berkeley, CA.
Clark, H. (1994) 'Taking up space: redefining political legitimacy in New York City', *Environment and Planning A*, 26(6): 937–955.

Clark, A. (2009) 'From neighbourhood to network: a review of the significance of neighbourhood in studies of social relations', *Geography Compass*, 3(4): 1559–1578.

Cope, M. (2008) 'Patchwork neighborhood: children's urban geographies in Buffalo, New York', *Environment and Planning A*, 40(12): 2845–2863.

Dikeç, M. (2005) 'Space, politics, and the political', *Environment and Planning D: Society and Space*, 23(2): 171–188.

Docherty, I., Goodlad, R. and Paddison, R. (2001) 'Civic culture, community and citizen participation in contrasting neighbourhoods', *Urban Studies*, 38(12): 2225–2250.

Ellaway, A., Macintyre, S. and Kearns, A. (2001) 'Perceptions of place and health in socially contrasting neighbourhoods', *Urban Studies*, 38(12): 2299–2316.

Elwood, S. (2006) 'Beyond cooptation or resistance: urban spatial politics, community organizations, and GIS-based narratives', *Annals of the Association of American Geographers*, 96(2): 323–341.

Elwood, S. and Leitner, H. (2003) 'Community-based planning and GIS: aligning neighborhood organizations with state priorities?', *Journal of Urban Affairs*, 25(2): 139–157.

Escobar, A. (2001) 'Culture sits in places: reflections on globalism and subaltern strategies of localization', *Political Geography*, 20(2): 139–174.

Forrest, R. and Kearns, A. (2001) 'Social cohesion, social capital and the neighbourhood', *Urban Studies*, 38(12): 2125–2143.

Galster, G. C. (1986) 'What is neighborhood? An externality-space approach', *International Journal of Urban and Regional Research*, 10(2): 243–263.

Galster, G. C. (2001) 'On the nature of neighbourhood', *Urban Studies*, 38(12): 2111–2124.

Hankins, K. (2007) 'The final frontier: charter schools as new community institutions of gentrification', *Urban Geography*, 28(2): 113–128.

Hankins, K. and Walter, A. (2012) ' "Gentrification with justice": an urban ministry collective and the practice of place-making in Atlanta's inner city neighborhoods', *Urban Studies*, 49(7): 1507–1526.

Harvey, David (1996) *Justice, Nature, and the Geography of Difference*. Blackwell Publishing: Malden, MA.

Hunter, A. (1979) 'The urban neighborhood: its analytical and social contexts', *Urban Affairs Quarterly*, 14(3): 267–288.

Jessop, B., Brenner, N. and Jones, M. (2008) 'Theorizing socio-spatial relations', *Environment and Planning D: Society and Space*, 26(3): 389–401.

Kearns, A. and Parkinson, M. (2001) 'The significance of neighbourhood', *Urban Studies*, 38(12): 2103–2110.

Laclau, Ernesto (1990) *New Reflections on the Revolution of Our Time*. Verso: London.

Leitner, H., Sheppard, E. and Sziarto, K. (2008) 'The spatialities of contentious politics', *Transactions of the Institute of British Geographers*, 33(2): 157–172.

Lupton, Robert (1997) *Return Flight: Community Development Through Reneighboring Our Cities*. FCS Ministries: Atlanta, GA.

Martin, D. (2003a) 'Enacting neighborhood', *Urban Geography*, 24(5): 361–385.

Martin, D. (2003b) 'Place-framing and place-making: constituting a neighborhood for organizing and activism', *Annals of the Association of American Geographers*, 93(3): 730–750.

Martin, D. G., McCann, E. and Purcell, M. (2003) 'Space, scale, governance, and representation: Contemporary geographical perspectives on urban politics and policy', *Journal of Urban Affairs* 25(2): 113–121.

Martin, L. (2007) 'Fighting for control: political displacement in Atlanta's gentrifying neighborhoods', *Urban Affairs Review*, 42(5): 603–628.

Massey, D. (1991) 'A global sense of place', *Marxism Today*, 35: 24–29.

Massey, D. (1994) *Space, Place, and Gender*. University of Minnesota Press: Minneapolis, MN.

Massey, D. (2004) 'Geographies of responsibility', *Geografiska Annaler*, 86B(1): 5–18.

Massey, D. (2005) *For Space*. Sage: London.

McCann, E. J. (1999) 'Race, protest, and public space: contextualizing Lefebvre in the U.S. city', *Antipode*, 31(2): 163–184.

McCann, E. J. (2001) 'Collaborative visioning or urban planning as therapy? The politics of public–private policy making', *Professional Geographer*, 53(2): 207–218.

McCann, E. J. (2003) 'Framing space and time in the city: Urban politics, urban policy, and the politics of scale', *Journal of Urban Affairs*, 25(2): 159–178.

McCann, Eugene and Ward, Kevin (2011) *Mobile Urbanism*, University of Minnesota Press: Minneapolis, MN.

McCarthy, John and Zald, Mayer (1973) *The Trend of Social Movements*. General Learning: Morristown, NJ.

McCarthy, John and Zald, Mayer (1977) 'Resource mobilization and social movements', *American Journal of Sociology*, 82(6): 1212–1241.

Meegan, R. and Mitchell, A. (2001) '"It's not community round here, it's neighbourhood": neighbourhood change and cohesion in urban regeneration policies' *Urban Studies*, 38(2): 2167–2194.

Merrifield, A. (1993) 'Place and space: a Lefebvrian reconciliation', *Transactions of the Institute of British Geographers*, 18(4): 516–531.

Mumford, Lewis (1968) *The Urban Prospect*. Harcourt, Brace & World: New York.

Newman, K. and Ashton, P. (2004) 'Neoliberal urban policy and new paths of neighborhood change in the American inner city', *Environment and Planning A*, 36(7): 1151–1172.

Olson, P. (1982) 'Urban neighborhood research: its development and current focus', *Urban Affairs Quarterly*, 17(4): 491–518.
Park, Robert, Burgess, Ernest and McKenzie, Roderick (eds) (1967[1925]) *The City*. University of Chicago Press: Chicago, IL.
Perkins, John (1995) *Restoring At-Risk Communities: Doing It Together and Doing It Right*. Baker Books: Grand Rapids, MI.
Pierce, J., Martin, D. and Murphy, J. (2011) 'Relational place-making: the networked politics of place', *Transactions of the Institute of British Geographers*, 36(1): 54–70.
Purcell, Mark (2008) *Recapturing Democracy: Neoliberalization and the Struggle for Alternative Urban Futures*. Routledge: New York.
Purdue, D. (2001) 'Neighbourhood governance: leadership, trust and social capital', *Urban Studies*, 38(12): 2211–2224.
Raco, M. (2000) 'Assessing community participation in local economic development – Lessons for the new urban policy', *Political Geography*, 19(5): 573–599.
Raco, M. and Flint, J. (2001) 'Communities, places and institutional relations: assessing the role of area-based community representation in local governance', *Political Geography*, 20(5): 585–612.
Rancière, Jacques (1999) *Dis-agreement: Politics and Philosophy*, trans. J Rose. University of Minnesota Press: Minneapolis, MN.
Rancière, J. (2001) 'Ten theses on politics', *Theory and Event*, 5(3) 17–34.
Robinson, A. (2001) 'Framing Corkerhill: identity, agency, and injustice', *Environment and Planning D: Society and Space*, 19(1): 81–101.
Schmidt, D. (2008) 'The practices and process of neighborhood: the (re)production of Riverwest, Milwaukee, Wisconsin', *Urban Geography*, 29(5): 473–495.
Sheppard, E. (2002) 'The spaces and times of globalization: place, scale, networks, and positionality', *Economic Geography*, 78(3): 307–330.
Suttles, G. D. (1972) *The Social Construction of Communities*. University of Chicago Press: Chicago, IL.
Traugott, M. (1978) 'Reconceiving social movements', *Social Problems*, 26(1): 38–49.

3

SPLINTERED GOVERNANCE: URBAN POLITICS IN THE TWENTY-FIRST CENTURY

Kevin Ward

> The complexity and richness of urban politics cannot be reduced either to the simplicities of collective consumption or those of growth coalitions. (Cochrane, 1999: 124)

> What is commonly defined as 'urban politics' is typically quite heterogeneous and by no means referable to struggles with, or among, the agents structured by some set of social relations corresponding unambiguously to the urban. (Cox, 2001: 756)

Introduction

Are you sitting somewhere in a city as you begin to read this chapter? If so, look out of your window. What do you see? The chances are you see buildings, and lots of them. Tall and short ones, old and new ones: the urban built environment embodies the past contests and struggles of architectures, consultants, developers, economists, engineers, land owners, planners and politicians, as well as members of what might rather quaintly be labelled as the 'general public'. Disputes over when and what to build, or when or what to tear down are ones that have characterized most cities, and continue to do so. Of course, it's not just about the buildings. It is about the land on which they are built. As Harvey (1982) has reminded us, under the capitalist system land should be treated as a purely financial asset, one that has to become a form of fictitious capital. As he puts it, '[l]andowners can coerce or cooperate with capital to ensure the creation of enhanced ground rents in the future ... [and] [t]hey thereby shape the geographical structure of production, exchange and consumption' (Harvey 1989a: 96). It is in and through the political that this coercion and/or the co-operation – as well as much more – occurs. Various struggles take place between a range of stakeholders, over 'bigger' issues, such as land zoning and residential eviction as well as over 'smaller' issues such as the design and security of the built environment and infrastructure.

So, turn back to the window, and look out again. Don't just see the bright new shopping mall, the iconic skyscraper, the old, run-down and disused lot, or the slightly 'average' row of shops, interspersed with examples of street furniture such as bins and seats. Instead, reflect on the politics of how these buildings have come to be. Consider the struggles over their building and maintenance, over the political in the broadest sense, both formally involving elected political parties, and less formally, involving all those others whose involvement, whose actions or inactions, were behind the production of the built environment you now see. This is primarily about what you see, but it's also about what you do not – about alternative plans, strategies or visions that never came to be. That were considered too radical perhaps, or that emphasized use over exchange values. Or, about those places from which the city that you can see through your window has learnt. For example, think about those cities that have become models for particular ways of developing (and revalorizing) the built environment, such as Baltimore (and its waterfront), Barcelona (and its mixed use), Freiburg (and its sustainability) and New York City (and its quality of life campaigns) (Nasr and Volait, 2003; Ward, 2006). Unless you are in these cities reading this chapter then you won't literally be able to see them through your window. That does not mean though that their influence is not present in the built environment you can see. There is a long list of cities whose recent developmental past owes more than a little to the experiences in these 'iconic' examples. There is certainly a politics to comparison and learning that lies behind what you can see through your window and how it came about (Robinson, 2005; Peck and Theodore, 2010; Ward, 2010; McCann and Ward, 2011).

Building on these themes, this chapter is organized into three sections. It reviews and evaluates two bodies of work that have sought to theorize urban politics. The first concerns the set of literatures that emerged in the 1970s around the notion of 'collective consumption'. For some this has been cast as 'old' urban politics. This is followed by a discussion of what Cox (1993) termed the 'new urban politics'. In this work the emphasis is on urban development. While there is much that divides these two literatures, there is also much that unites them, as this section concludes. Most directly, both literatures rather take for granted what is meant by, and goes into the making up of, 'the urban'. This reflects a wider trend around emphasizing the territorial nature of 'urban' politics. There is nothing wrong with this, as work from this tradition continues to provide theoretically sophisticated accounts of the current urban political condition around the world.

However, the second section of this chapter turns to more recent attempts to theorize urban politics relationally. This has involved rethinking the where and what of urban politics. The chapter argues that one way of conceiving of the 'urban' in urban politics is through thinking about it in terms of the co-presence of splintered elements of elsewhere. That is how 'the urban' in urban politics is assembled, involving different types of labour. In conclusion the chapter makes two points: first, traditional theories of urban politics, while not without their insights, have tended to work with a rather bounded theorization of the 'urban'; and second, one fruitful way forward for work in this field would seem to be to understand the 'urban' as a place that gives shape to channels, networks and webs of differing geographical complexity and reach. This necessitates a revisiting of what and where goes into the theoretical understandings of 'urban politics' with which we work.

'Old' and 'new' urban politics reconsidered

Let's start with 'old' urban politics. *La Question Urbaine* by Manuel Castells was published in 1972. It was subsequently published in English as *The Urban Question: A Marxist Approach* and it spawned a series of studies that sought to 'ground urban politics in collective consumption' (Cox and Jonas, 1993: 10). In the initial formulation capital's failure to secure the reproduction of labour power was behind the state's intervention in the collective consumption of 'public' goods such as education, health and transport, all commodities whose production is not guaranteed by capital because they have little or no market price. In intervening in this way, the state sought to manage the needs of labour but in the interests of furthering capital accumulation. A politics around elements of economic and social life emerged, around everything from affordable housing to the cleaning of the streets, from the provision and maintenance of the public schools to the building of public playgrounds. As Merrifield (2002: 120) put it:

> By understanding collective consumption ... Marxists could better understand the latest dynamics of the city, could better understand social change, could better understand why capitalism was still around and what the pressure points were in its prevailing 'monopoly stage'. The urban structure, [Castells] believed, made more sense now.

Subsequent studies sought to develop and refine this framework. Most importantly, Peter Saunders (1986) devised the notion of the 'dual state' based on his housing fieldwork in the UK. This introduced the notion of a clear scalar division of labour of the state. The central state was represented as where economic decisions rested. It was here corporatist decision-making dominated infrastructural and inward investment. The pluralist local state – understood as a set of social relations rather that a set of institutions – oversaw consumption issues. The urban was the territorial arena in which struggles over how resources and spending were distributed to different sections of the population. Social movements of various sorts were involved in these struggles. Various studies pointed to the possibilities and the limits of constructing alternative approaches to capitalism from below.

While these issues of collective consumption, social reproduction, and the state continue to this day to frame the studying of urban politics (Jonas and Ward, 2007), the last three decades have also been witness to the emergence of what Cox (1993) terms the 'new urban politics' (NUP). This particular emphasis on the politics of economic development stemmed from a belief that capital had become more mobile, which had put localities and those that govern them, on the back foot. According to Cox (1995: 455): 'The NUP has two points. The first is that urban politics is about local economic development ... The second point refers to the framework for understanding this politics. Major significance is assigned to relations with more global events.'

More specifically, there are four features that have characterized this newer theorization of urban politics. These are its attention to *what* constitutes 'urban politics', *who* – the agents – that participate in it, *why* they are involved and *what* policies are pursued. Cox (1993: 433) put it thus: '[q]uite clearly, urban development, for many scholars, is now what the study of urban politics is about.' According to Cox (1995: 214) the substance of the NUP

> is ... the competition of local governments in metropolitan areas for major shopping malls; the resuscitation of downtowns through public investment in convention centres and enclosed shopping malls and through the promotion of gentrification; the competition cities for airline hubs as stimulus to attracting corporate headquarters; and conflicts over the funding of these projects.

Thus the NUP is about improving the business climate, 'that particular bundle of investments and activities, which has the best prospect for enhancing future rents' (Harvey, 1989a: 97).

In terms of the second feature, those that are involved in the NUP, much has been made of the ways in which the state, capital and other agents have worked in unison to oversee a transformation in the ways in which cities are governed (Logan and Molotch, 1987; Harvey, 1989b; Jonas and Wilson, 1999; Ward, 2000; Cochrane, 2007). Such was the close working that a point was reached 'where the local state ends and private firms begin is often very obscure' (Cox and Mair, 1988: 311). This has constituted nothing less than the qualitative restructuring of the state (Brenner, 2004), an argument that has been pushed further in recent years by work on the city and neo-liberalization (Brenner and Theodore, 2002; Wilson, 2004; Hackworth, 2007). Terms such as 'coalition' and 'regime' were used to invoke the structured participation of differently situated agents (Logan and Molotch, 1987; Harvey, 1989a; Stone, 1989). Of course these concepts have rather different theoretical lineages. These should not be overlooked. Nevertheless, what they all speak to is an attention to the kinds of territorial arrangements that were assembled in a number of cities over securing the conditions for future rents, as it has been termed by Harvey (1989a).

The third theme that runs through much of the NUP literature concerns different agents' material rationales for participation. Cox (1991: 273; see also Cox and Mair, 1988: 1991) developed the notion of 'local dependence'. This argued that 'all firms, branches of the state, are locally dependent, that is dependent on a localized set of social relations'. He questioned the assumption that capital is always mobile, which characterized much of the NUP work. Rather, the term 'local dependence' sought to capture the 'fixity' or 'embeddedness' of different interests and how this condition translates – or not – into a particular form of politics. As Cox (1995: 216) explained, 'the interest in the expansion of local economies arises from the existence of capitalist/worker/stage agency interests in the appropriation of profits/wages/taxes in particular places'.

The fourth and final characteristic of the NUP work has been that on the types of policies pursued by the different territorial alliances. While there is some evidence of diversity, there are also plenty to suggest that 'policies have the usual 'good business climate' menu of tax abatements, cutting 'red tape' and securing the necessary infrastructural investments' (Cox and Jonas, 1993: 12). So, stop for a minute, and look outside that window again. Can you see revalorized old warehouses or newly built retail malls, or, perhaps, swanky restaurants or newly refurbished public spaces? Where you are will determine what you see, of course. If you are sitting downtown it is likely that you'll be able to see one or more of these examples of how the built environment was restructured over the 1980s and 1990s. If you are in a neighbourhood beyond the centre you might be looking at a newly developed industrial park or refurbished high tech business incubator. That is not to say these alliances have in every city been successful, in terms of securing the conditions for capital accumulation. They have not. So, out of your window you may spot examples of where capital has not got its own way, where land remains underdeveloped, or underutilized. Empty lots, or run down retail outlets, perhaps.

Underpinning both the emphasis in the 'old urban politics' of collective consumption and social reproduction and in the 'new urban politics' of economic development is a quite specific understanding of territory. In much of these literatures the notion of the urban is left rather unchallenged. It is taken for granted, often, although not always, a shorthand for the geographical remit of local government. There are some exceptions, where a wider theoretical appreciation of the range of geographical scales present in what in shorthand form it termed 'the urban'. For example, Harvey (1989b: 6) and Cox (1998: 19) both sort to emphasize the different scales represented in 'the urban':

> The shift towards entrepreneurialism in urban governance has to be examined ... at a variety of spatial scales – local neighbourhood and community, central city and suburb, metropolitan, region, nation state, and the like.
> 'Local politics appears as metropolitan, regional, national, or even international as different organizations try to secure those networks of associations through which respective projects can be realized.'

So, there is a tradition of sorts of allowing for the multiplicity of scales that are present in any single 'scale' – local, regional, and so on. However, the ontological purchase of the scalar vocabulary remains undiminished (Brenner, 2004: Jonas, 2006). Territories, in the form of 'the urban' or 'the national' remain central to the explaining of urban politics. The next section, however, introduces another means of theorizing urban politics, which I argue is complementary to the more traditional work discussed thus far.

Splintered 'Urban' politics

> [a]n extrospective, reflexive, and aggressive posture on the part of local elites and states ... Today cities must actively – and responsively – scan the horizon for investment and promotion opportunities, monitoring 'competitors' and emulating 'best practice', lest they be left behind in this intensifying competitive struggle for the kinds of resources (public and private) that neoliberalism has helped make (more) mobile. (Peck and Tickell, 2002: 47)
> So 'local' policy development now occurs in a self-consciously comparative and asymmetrically relativized context. The boundaries of local jurisdictions and policy regimes would seem, therefore, to be rather more porous than before. (Peck, 2003: 229)

Recent years have seen a series of literatures emerge that have challenged us all to re-think the language of geographical scale and territory. This is reflective of a larger debate between those who propose different ways of conceiving space and it has highlighted the increasingly open, porous and inter-connected configuration of territorial entities (Massey, 1993; Amin, 2004). Amin (2007: 103) argues for the need to consider cities as sites 'of intersection between network topologies and territorial legacies'. For Allen and Cochrane (2007) it is important to appreciate both the relational as well as the territorial elements bound up in 'the urban' and 'the regional' politics of economic development. There are elements of elsewhere that go into the assembling of 'urban politics', such as other cities' experiences, models and success stories. Or, viewed from one particular city – such as the one in which you are located – urban politics often seems to be about bringing together 'splinters' of other cities. That is 'splinters' or aspects of what might be termed as approaches, models, policies, programmes or strategies. In each city these different 'splinters' are reassembled to take specific territorial forms.

In my own city, which I see when I look out of my window at work, I can see evidence of Manchester comparing, imitating and referencing Barcelona. Sometimes this is quite obvious, as in Manchester's Catalan Square. In other ways it is more subtle, as in Manchester's attempt to pick and mix aspects of the larger 'Barcelona model' of urban regeneration. And, of course, Manchester is not alone in doing this, and drawing on this city and its experiences (García-Ramón and Albet, 2000). There is more and more evidence of cities engaging in policy tourism, visiting other cities to see for themselves what can be achieved, or hosting visits, in order to learn from visiting delegates (Ward, 2011).

The burgeoning academic literature that understands urban politics as being about more than what happens within the city limits (Ward, 2006, 2007; Peck and Theodore, 2010; McCann 2011; McCann and Ward, 2011), 'the urban' in understood as 'both a place (a site or territory) and as a series of unbounded, relatively disconnected and dispersed, perhaps sprawling activities, made in and through many different kinds of networks stretching far beyond the physical extent of the city' (Robinson, 2005: 763).

One of the means through which what goes in one city is made to matter to developments in another city is via the notion of comparison. Through this apparent relational proximity can be manufactured a sense of physical proximity. Examples of policies or strategies in a city are turned into models that can then be made mobile, moved from one city to another. Comparison, as Peck (2003) suggests at the start of this section, is now an important aspect of urban politics. And, of course, the comparison of today has some past precedents (Nasr and Volait, 2003; Healey and Upton, 2010). However, recent years appear to have seen comparisons made more quickly and with greater effect for those being compared. Put another way, the geographical reach of the reference points appears to have been extended. And comparison begets comparison, as there is an urban politics over which city to be compared with and on what terms, with different stakeholders choosing different places of comparison, and in the process, opening up for consideration the different 'wheres' that go into the making up of 'urban' politics.

Calgary and the politics over its Plan It Calgary process is an interesting example of how contemporary urban politics consists of both territorial and relational elements. A Canadian city with a population of just under one million, in 2005 the city government consulted with almost twenty thousand of its residents in an exercise called imagineCALGARY (see www.imaginecalgary.ca/). The purpose was to begin to consider what sort of future the city might have. In 2006 a document was produced – 'City owned – community initiative' (City of Calgary, 2006: n.p.). This established a one hundred year vision. This was an interesting document. Various territorial stakeholders were involved in its production and discussion. Building on this, in 2007 the City government began to develop an integrated Municipal Development Plan (MDP) and Calgary Transportation Plan (CTP). To ensure that these plans resonated with imagineCALGARY the city government devised Plan It Calgary. Its purpose was

> To set out a long-term direction for sustainable growth to accommodate another 1.3 million people over the next 60 years. It was grounded in the values of SMART growth and Council adopted sustainability principles for land use and mobility. These principles focused on a compact city form that cultivates walking, cycling and transit, and preserves open space, parks and other environmental amenities. (www.calgary.ca/planit/)

The debate over the Plan It Calgary process involved all manner of Calgary-based stakeholders, as well as others whose geographical reach extended beyond the city. Examples included local businesses, the local media and the local university, in addition to city government. So, it would be possible to draw on traditional territorial understandings of 'the urban' to analyse this case. Pro-growth versus anti-growth coalitions, fighting it out over the future of the city of Calgary, in the process arguing over what constitutes 'growth'. However, the urban politics around Plan It Calgary reached out beyond the city limits. Various other places were present in the politics around the scheme. A couple of examples will suffice.

Some homebuilders and developer groups in the city did not like the arguments that were being made for a denser, more transit-centric Calgary. They were unhappy that Calgary was being implored to compare itself with the likes of Portland, Oregon, which has become somewhere that other cities can emulate in terms of its sustainable and smart growth stance. These groups brought in Randal O'Toole of the Cato Institute to discuss the limits to what was being suggested. He, of course, came with his own ideological baggage – working at the Cato Institute and having established the antiplanner.com website. He did not let down those who involved him. He sharply criticized the Smart Growth principles of Denver and Portland in the USA and Toronto and Vancouver in Canada – precisely those cities against which Calgary was beginning to benchmark itself. He argued that those other cities have adopted policies that have caused soaring house prices and excessively subsidized public transit. In turn, he traded in one set of comparisons for another. He argued that Calgary should be comparing itself with, and in the process imitating, Houston in the USA. This was a city with a very different track record on sustainability to those referenced in debates in Calgary, such as Portland. These references to elsewhere, of seeking to bring another city's experiences to bear, was an important element in the 'urban' politics around the Plan It Calgary process.

By the end of 2009 the Plan It Calgary process had delivered a new Municipal Development Plan (MDP) and Calgary Transportation Plan (CTP). These established the parameters for the city's growth and development into the middle end of the twenty-first century. This did not mark the end of the territorial and relational politics. Just recently Miller (2009: n.p.) has argued that:

> As Calgary struggles toward becoming a more sustainable city, there's much it can learn from cities like Freiburg. Calgary certainly can't remake itself in the image of Freiburg – we have to start from where we are and create our own future. But by the same token, we don't have to re-invent every wheel. There's much we can learn from sustainability initiatives around the world. (www.facebook.com/note.php?note_id=91014104693)

The socio-spatial struggles within cities over particular developmental pathways and trajectories are ones in which comparisons are made and re-made to other localities. The Plan It Calgary example highlights how other cities became bound up in each other's attempts to develop future strategies, revealing how cities – 'as local alliances attempting to create and realize new powers to intervene in processes of geographical restructuring' (Cox and Mair, 1991: 208) – do contemporary urban politics.

Conclusion

This chapter has presented an overview of the changing ways in which urban politics has been studied. Stemming from the seminal work of Castells (1977) and Harvey (1982), much of the work in human geography has tended to take the urban for granted. Greater attention has been paid to the substance of urban politics. From the emphasis on collective consumption through to the work on economic development, studies have focused on what it is that is being struggled over. In the latter it tended to be over education, health, social services and transport, while in the former it tended to be over inward investment, physical regeneration and residential gentrification, this difference in Harvey (1989b) being accompanied by a shift in its management.

Most recently work that builds on that produced under the 'new urban politics' (NUP) banner has been part of a wider literature that has begun to pay serious attention to what and where goes into the making-up of the urban in urban politics. Taking its intellectual leave from inroads made in other aspects of human geography, this research has sought to re-conceive 'urban politics' as comprising different (often territorial) elements of elsewhere, brought and held together by various agents and given a particular territorial expression. In particular, a strand of work has emerged that has sought to complement previous work that emphasized the territorial nature of urban politics. This more recent literature instead has strived to emphasize the means in which cities are interconnected through the movement of models and policies from one place to another and what this means for territorial understandings of urban politics (McCann and Ward, 2011).

So, you might ask, returning to your window, what does this means for what you can see? Well, it would seem to mean two things conceptually: first, the nature of the social relations *between* localities warrant further study. Which agents are involved in the movement – borrowing, evaluation, learning, etc. – of policies from one place to another and how are the processes of comparison and learning performed? Second, the material and territorial outcomes in one place matter to what happens in another place. Localities are implicated in each other's pasts, presents and futures in a way that challenges more traditional ways of conceiving of 'urban politics'. So, the territorial and the relational are intertwined in the urban politics of here and there, in the process, troubling the where of 'here' and of 'there'. Now there is something to mull over when you turn back to your window.

References

Allen, J. and Cochrane, A. (2007) 'Beyond the territorial fix: regional assemblages, politics and power', *Regional Studies*, 41(9): 1161–1175.

Amin, A. (2004) 'Regions unbound: towards a new politics of place', *Geografiska Annaler*, 86B: 33–44.

Amin, A. (2007) 'Re-thinking the urban social', *City*, 11(1): 100–114.

Brenner, Neil (2004) *New State Spaces: Urban Governance and the Re-scaling of Statehood*. Oxford University Press: Oxford.

Brenner, N. and Theodore, N. (2002) 'Preface: from the "new localism" to the spaces of neoliberalism', *Antipode*, 34(3): 341–347.

Castells, Manuel (1977) *The Urban Question: A Marxist Approach*. MIT Press: Cambridge, MA.

City of Calgary (2006) *Imagine Calgary Plan for Long Range Urban Sustainability*. City of Calgary: Calgary.

Cochrane, Allan (1999) 'Redefining urban politics for the twenty first century', in Jonas, A.E.G. and Wilson, D. (eds) *The Urban Growth Machine: Critical Perspectives Two Decades Later*. State University of New York: Albany, NY. pp. 109–124.

Cochrane, Allan (2007) *Understanding Urban Policy: A Critical Approach*. Blackwell: Oxford.

Cox, K.R. (1991) 'Questions of abstraction in studies in the New Urban Politics', *Journal of Urban Affairs*, 13(3): 267–280.

Cox, K. R. (1993) 'The local and the global in the new urban politics: a critical view', *Environment and Planning D: Society and Space*, 11(4): 433–448.

Cox, K. R. (1995) 'Globalisation, competition and the politics of local economic development', *Urban Studies*, 32(2): 213–224.

Cox, K. R. (1998) 'Spaces of dependence, spaces of engagements and the politics of scale, or: looking for local politics', *Political Geography*, 17(1): 1–23.

Cox, K. R. (2001) 'Territoriality, politics and the "urban"', *Political Geography*, 20(6): 745–762.

Cox, K. R. and Jonas, A. E. G. (1993) 'Urban development, collective consumption and the politics of metropolitan fragmentation', *Political Geography*, 12(1): 8–37.

Cox, K. R. and Mair, A. (1988) 'Locality and community in the politics of local economic development', *Annals of the Association of American Geographers*, 78(2): 307–325.

Cox, K. R. and Mair, A. (1991) 'From localized social structures to localities as agents', *Environment and Planning A*, 23(2): 197–213.

García-Ramón, M.-D. and Albet, A. (2000) 'Commentary: pre-Olympic and post-Olympic Barcelona: a "model" for urban regeneration today?', *Environment and Planning A*, 32(8): 1331–1334.

Hackworth, Jason (2005) *The Neoliberal City*. Cornell University Press: Ithaca, NY.

Hackworth, J. (2007) *The Neoliberal City: Governance, Ideology and Development in American Urbanism*, Cornell University Press, Ithaca, NY.

Harvey, David (1982) *The Limits to Capital*. Verso: London.

Harvey, D. (1985) 'The geopolitics of capitalism', in Gregory, D. and Urry, J. (eds) *Social Relations and Spatial Structures*. Macmillan: London. pp. 128–163.

Harvey, David (1989a) *The Urban Experience*. Baltimore: The John Hopkins University Press.

Harvey, D. (1989b) 'From managerialism to entrepreneurialism: the transformation in urban governance in late capitalism', *Geografiska Annaler*, 71B: 3–17.

Healey, P. and Upton, R. (eds) (2010) *Crossing Borders: International Exchange and Planning Practices*. London: Routledge.

Hoyt, L. (2006) 'Importing ideas: the transnational transfer of urban revitalization policy', *International Journal of Public Administration*, 29: 221–243.

Jessop, B., Peck, J. and Tickell, A. (1999) 'Retooling the machine: economic crisis, state restructuring, and urban politics', in Jonas, A. E. G. and Wilson, D. (eds) *The Urban Growth Machine: Critical Perspectives Two Decades Later*. State University of New York Press: Albany, NY. pp. 141–159.

Jonas, A. E. G. (2006) 'Pro-scale: further reflections on the "scale debate" in human geography', *Transactions of the Institute of British Geographers*, 31(3): 399–406.

Jonas, A. E. G. and Ward, K. (2007) 'Introduction to a debate on city regions: new geographies of governance, democracy and social reproduction', *International Journal of Urban and Regional Research*, 31(1): 169–178.

Jonas, A.E.G. and Wilson, D. (eds) (1999) *The Urban Growth Machine: Critical Perspectives Two Decades Later*. State University of New York Press: Albany, NY.

Logan, John R. and Molotch, Harvey (1987) *Urban Fortunes: The Political Economy of Place*. University of California Press, CA: Berkeley.

Massey, D. (1993) 'Power-geometry and a progressive sense of place', in Bird J., Curtis, B., Putman, T., Robertson, G. and Tickner, L. (eds) *Mapping the Futures: Local Cultures, Global Change*. Routledge: London. pp. 59–69.

McCann, E. (2007) 'Expertise, truth, and urban policy mobilities: global circuit of knowledge in the development of Vancouver, Canada's "four pillar" drug strategy', *Environment and Planning A*, 40(4): 885–904.

McCann, E. (2011) 'Urban policy mobilities and global circuits of knowledge: towards a research agenda', *Annals of the Association of American Geographers*, 101(1): 107–130.

McCann, E. and Ward, K. (2011) *Mobile Urbanism: Cities and Policymaking in the Global Age*. University of Minnesota Press, MN: Minneapolis.

Merrifield, Andy (2002) *MetroMarxism: A Marxist Tale of the City*. Routledge: New York.

Miller, B. (2009) *Inspiration for a sustainable city*, Byron Miller's Dispatches from Freiburg, 22 June, available at: https://www.facebook.com/note.php?note_id=91014104693.

Nasr, J. and Volait, M. (eds) (2003) *Urbanism Imported or Exported? Native Aspirations and Foreign Plans*. Wiley Academy: Chichester.

Peck, J. (2003) 'Geography and public policy: mapping the penal state', *Progress in Human Geography*, 27(2): 222–231.

Peck, J. and Theodore, N. (2010) 'Mobilizing policy, models, methods and mutations', *Geoforum*, 41(2): 169–174.

Peck, J. and Tickell, A. (2002) 'Neoliberalizing space', *Antipode* 34(3): 380–404.

Prince, R. (2010) 'Policy transfer as policy assemblage: making policy for the creative industries in New Zealand', *Environment and Planning A*, 42(1): 169–186.

Robinson, J. (2005) *Ordinary City: Between Modernity and Development*. Routledge: London.

Robinson, J. (2008) 'Developing ordinary cities: city visioning processes in Durban and Johannesburg', *Environment and Planning A*, 40(1): 74–87.

Saunders, Peter (1986) *Social Theory and the Urban Question*. Routledge: London.

Stone, Clarence (1989) *Regime Politics: Governing Atlanta 1946–1988*. University Press of Kansas Press: Lawrence, KS.

Ward, K. (2000) 'A critique in search of a corpus: re-visiting governance and re-interpreting urban politics', *Transactions of the Institute of British Geographers*, 25(2): 169–185.

Ward, K. (2006) '"Policies in motion", urban management and state restructuring: the trans-local expansion of Business Improvement Districts', *International Journal of Urban and Regional Research*, 30(1): 54–75.

Ward, K. (2007) 'Business Improvement Districts: policy origins, mobile policies and urban liveability', *Geography Compass*, 2(3): 657–672.

Ward, K. (2010) 'Towards a relational comparative approach to the study of cities', *Progress in Human Geography*, 34(3): 471–487.

Ward, K. (2011) 'Entrepreneurial urbanism, policy tourism and the making mobile of policies', in Bridge, G. and Watson, S. (eds) *The New Companion to the City*. Wiley-Blackwell: Oxford.

Wilson, D. (2004) 'Towards a contingent urban neoliberalism', *Urban Geography* 25(8): 771–783.

4

RUTHLESS: THE FORECLOSURE OF AMERICAN URBAN POLITICS

Elvin Wyly and Kathe Newman

Urban Politics, Refinanced

Urban politics is defined by the scale and location of power dynamics within urban areas. When Dahl (1961: v) asked the 'ancient question' of who governs, he offered answers 'by examining a single American city in New England', specifically New Haven, Connecticut. Similarly, urban regime theory (e.g. Stone, 1993: 1) focuses on the creation and maintenance of governing capacity through 'the creative exercise of political choice', involving the mobilization of coalition partners 'to craft arrangements through which resources can be mobilized' to achieve the goals of an urban community. Likewise, the enormous and influential literature inspired by Harvey Molotch's powerful 'growth machine' metaphor (see Jonas and Wilson, 1999) has documented the many ways that rentiers in the city maintain (local) political power as they extract tributes and capital from the city. With few exceptions, urban politics still involves the study *in* and *of* the urban in an attempt to understand the politics.

Yet it is now widely acknowledged that the intensified financialization of post-industrial economic processes has dramatically rescaled urban politics. The massive housing and credit boom of the first decade of the twentieth century – and the subsequent global crisis – require us to consider the specific contribution of new scales of housing finance. Put simply, the economics and politics of mortgage finance may have been decisive in shifting the locus of urban politics beyond the urban realm – and into explicitly *anti-urban* political venues. Orderud (2011: 1218) observed, 'the linkage between local capital and local property was loosened ... and finance became an active player in forming the urban landscape on the basis of a capitalist logic characterized by cyclical periods of accumulation'. Particularly for heavily deindustrialized cities, financialized housing markets created new relations between capital and community – requiring new thinking about what produces the city and the scale of urban politics (Cox, 1993; DeFilippis, 1999; Martin, 2011). In this chapter, we work from the ground up in an attempt to understand the relations between the post-industrial city and the transnational networks of America's financialized housing economy. Local imprints of the housing boom and bust offer clues to the changing relations between capital and place, with major implications for local politics.

Historically, research about foreclosure focuses on the demand side of the equation by trying to predict why borrowers default. But another body of literature looks at the problem and wonders about supply. We make the case for understanding the systemic nature of finance capitalism and its relationship to the urban and to the institutions, rules, and relations that evolved to produce the catastrophic system failures of the worst financial crash since the Great Depression. Since the property and people associated with mortgages, default, and foreclosure are located in places, we believe that it is only through explicitly making these connections that we can untangle the financial crisis and its connection to the urban. Significantly, we argue that the spatial concentration of foreclosures reflects the structural dynamics of banking and finance rather than the flaws of individual borrowers or institutions. We agree with Levitan and Wachter's (2010) assessment that the credit boom must be understood in terms of supply-side transformations with serious consequences for homeowners and localities, and which, in subtle ways, have re-scaled urbanization. Responding to these changes requires an adaptive, agile scale politics to challenge the legal and institutional arrangements of exploitative capital accumulation.

In this chapter, we explore fine-grained geographies of foreclosure, test for explicit spatial relations in mortgage defaults, and analyse their relations to the contemporary infrastructure of securitization and capital market investment. We review the prevailing economic interpretation of mortgage foreclosure and draw on an interdisciplinary challenge that provides a compelling alternative. We then describe a research project to map defaults in one American metropolitan area, before developing and testing several hypotheses regarding the local-yet-transnationalized risks exacerbated by the subprime lending boom of 2002–2007. The results suggest a reorientation of our understanding of urban politics and a need to consider policies that address the problems that produced the global but very locally felt crisis.

Ruthless Default

Economists have long sought to understand how, why, and when borrowers default. The answers inform how credit is offered and priced which, in turn, affects who gets credit, how much it costs, and what risks it entails. Much of the mortgage default literature treats the mortgage as if it were like any other tradable commodity which allows the mortgage to be modelled with options theory (Ambrose and Capone, 1996). From the borrower's perspective, each payment due date presents three options: 1) remit the agreed payment; 2) pre-pay the entire principal balance (the call option); and 3) default (the put option). The call option is 'in the money' if interest rates fall and a borrower has sufficient resources (from non-housing wealth or from another lender) to retire high-cost debt with lower-cost credit. The put option is a rational choice if the market value of the home falls below the outstanding balance on the debt – that is, if the borrower has negative equity. Default is said to be *ruthless* when borrowers default when equity turns negative, as soon as the put option is in the money or 'far' in the money (Avery et al., 1996; Deng et al., 1996). When transaction costs such as the love of home, moving expenses, access to credit, and the reputational penalties of tarnished credit play little or no role in the decision to exercise the put option, default is said to be *frictionless*. Adding transaction costs to default prediction models produces impossibly low default rates leading some to reject them; others conclude that 'the value of the option will vary across people' which recognizes their importance and that they are hard to model (Quigley and Van Order, 1995: 113). Alternatively, the love of home or the costs of moving may keep borrowers from defaulting until confronted with a 'trigger event' like job loss or divorce; other borrowers may use money from other sources (Foote et al., 2008; Elmer and Seelig, 1999). Early default models did not include credit history, presumably one of the more important predictors of default, because it was hard to measure. But since the mid-1990s, technological changes including credit scoring enabled lenders to more finely assess risk, originate more loans quickly, and shift to computer underwriting which made entirely new forms of lending possible (Straka, 2000).

The conventional wisdom of options theory from Wall Street and public policy in Washington, DC held that spatial variations in mortgage termination were understood in terms of demand-side factors, especially the financial and personal circumstances that influence borrowers' debt burden and net equity position, as mediated by exogenous factors such as interest rates, housing values, and 'trigger events' like regional recessions and spikes in unemployment. The most potent 'recipe for delinquency involves young loans to low credit score borrowers with low or no documentation in housing markets with moderately volatile and flat or declining nominal house prices' (Danis and Pennington-Cross, 2008: 67). This profile is precisely the niche filled by the subsidized insurance programmes of the Federal Housing Administration. Accordingly, for many years, one of the most urgent policy questions was how options theory could predict default rates among low- and moderate-income borrowers, many of them racial and ethnic minorities with blemished credit, buying homes in low- or moderate-income neighbourhoods and in declining industrial cities (Foster and Van Order, 1985;

Vandell, 1993; Ambrose et al., 1997). Risk-based pricing made capital available to many who had lacked it. Beginning in the mid-1990s, the rapid growth of subprime lending reshaped urban areas and the debate over the meaning of geographically concentrated mortgage defaults.

Parallax Capital

For many years, a rigorous and compelling body of theoretical and empirical research challenged the orthodox interpretations of options theory and risk-based pricing (Dymski, 1999, 2009; Bunce et al., 2001; Engel and McCoy, 2002; Squires, 2003; Apgar et al., 2004; Immergluck, 2004, 2009a; Ashton, 2009; Wachter, 2009). This is a broad, diverse, and interdisciplinary literature, and its roots come from heterodox economics, law and economics, public policy, and contemporary urban political economy. Our central concern is with alternative understandings of geographically concentrated delinquency, default, and foreclosure that emerge from this critical literature.

Critical analysts draw attention to a constellation of processes summarized as *the urban problematic* – 'the complex, racialized dynamics of social inequality in urban space' (Dymski, 2009: 427). Three specific elements are fundamental to an understanding of today's foreclosure wave. First, changes in financial services, transnational capital markets, and state policy shifted the balance between demand and supply for mortgage credit. The expansion of the secondary market elevated mortgage-backed securities and derivatives from their position as obscure policy mechanisms of the 1980s to a higher status as attractive, high-yielding investments for global institutional investors. By the time budget surpluses led the U.S. Treasury to retire the 30-year long bond in 2001, the mortgage securities of the Government Sponsored Enterprises were sufficiently widely traded to become the global benchmark for evaluating the yield and risk profile of long-term debt commitments. Meanwhile, public policy shifts altered the historic advantages deposit-taking banks and thrifts enjoyed, and created advantages for differently regulated non-bank lenders which tapped into world financial markets in search of high yields in a climate of low inflation and interest rates (Gotham, 2009; Immergluck, 2004, 2009a). The result was to reorient demand and supply. We might then view foreclosures as less a reflection of demand-side variations in consumer risk profiles and, instead, more the outcome of supply-side imperatives of industry strategies and investment profiles (US GAO, 2007; Mian and Sufi, 2009; Levitan and Wachter, 2011). On the ground, these transformations were visible as lenders aggressively sought borrowers by creating and marketing loans that would tap new markets. Exotic loans captured borrowers in some markets while loose underwriting incorporated borrowers who would have been left out of others.

Second, local mortgage markets became more sharply divided. Apgar et al. (2004: 1) summarize the shift this way: 'With new low downpayment programs and a highly automated mortgage delivery system, the mortgage industry – often operating through a network of mortgage brokers – has dramatically expanded lending in the same low-income, low-wealth and minority neighborhoods that were once victimized by mortgage "redlining."' The technological advances of automated underwriting that were expected to weaken the significance of space and place had the opposite effect. A specialized industry of real estate data mining allowed lenders to gain access to customized, targeted prospect lists, often with borrowers' names, addresses, ages, current loan balances, measures of 'available equity', and, in the case of at least one firm, data on the prospect's 'ethnicity' (Immergluck, 2004, 2009a: 84).

A third element of the urban problematic added further complications through *institutional and legal rescaling*. Continuing a trend towards consolidation, first documented in the 1990s (Dymski, 1999), the mortgage business became more concentrated among the largest firms. The market share of the top twenty-five US lenders rose from about 30 per cent in 1992 to 78 per cent in 2002; for the top five subprime lenders, the share doubled from 20 per cent in 1996 to 40 per cent in 2002 (Apgar et al., 2004). Simultaneously, a long-running stalemate over federal regulation of abusive mortgage practices led many state legislatures to enact anti-predatory lending statutes. Federal agencies fought to pre-empt state restrictions on federally chartered institutions; other laws from the 1980s eliminated state usury rules on some loan products (Engel and McCoy, 2007). The result was a complex mosaic of loan products, made by traditional, closely supervised banks and savings institutions alongside less regulated mortgage companies which often tapped credit lines from investment banks and originated loans via virtually unregulated mortgage brokers. The legal environment provided few federal restrictions, and weak state rules that applied to small shares of the total market.

These changes merged during the housing and lending boom that took off after the short recession of 2001 and collapsed in 2007 and 2008, into the worst financial crisis since the Great Depression. These changes are captured in the notion of parallax capital – from the Greek *para* (beside, beyond) and *allasein*, to change. We hypothesize that the reconfiguration of American mortgage markets directly affected urbanization and set the stage for a geography of delinquency, default, and foreclosure. While access to capital initially appeared advantageous and many cities celebrated its return, the value was illusory. The effects reoriented risk in historically disinvested urban communities. These transformations set in place the need to reorient the way we think about urban change. These changes in the post-industrial city suggest the need to rescale urban politics to consider the broader forces shaping the city in a post industrialized financialized economy. This, in turn, suggests the need to construct an urban politics of finance that illuminates the political economy of urbanization.

In the rest of this chapter, we use the framework of an urban problematic reshaped by parallax capital to analyse the wave of mortgage defaults and foreclosures, focusing on the case of Essex County, New Jersey. Situated just west of New York City, Essex County includes neighbourhoods representative of some of the nation's wealthiest executive suburbs, some of its most distressed inner-ring, early twentieth-century suburbs, and some of its most disinvested inner-city landscapes. Looking from the ground up can help us to better understand the economic transformations that shape the city and reconsider an urban politics that responds to the source of these changes.

Locating Disaster

The first research step involved gaining access to public information on mortgage foreclosures, and translating the information into something useful for a local spatial analysis. New Jersey law specifies a judicial foreclosure process, in contrast to the quick, non-judicial actions permitted in some states. Information about the foreclosure process is located in foreclosure filings, the complaints filed at the state court house and in *Lis Pendens* notices which are filed in the county where the property is located. Addresses were geo-coded with 2008 Census TIGER files, which involved considerable efforts to clean address mistakes. We have information on about 15,695 foreclosure filings between 2004 and 2008. We were unable to assemble a complete dataset for 2005, excluded 3.5 per cent because of inadequate addresses, and removed 0.7 per cent for problems with origination or complaint dates.

Here we want to explore whether the foreclosure wave is associated with spatial neighbourhood effects. Our approach involves two steps. First, we measure the fine-grained local contours of the foreclosure wave, using those pieces of information that *can* be trusted as reliable, accurate, and meaningful: space (the address) and time (mortgage origination and foreclosure complaint dates). Patterns are tested for statistically significant spatial effects and different from random spatial variability, allowing the identification of zones of distinctive neighbourhood effects. These zones are used as input for the second step: testing whether the varied conditions of local markets during loan origination play a significant role in the emergence of spatial effects in loan defaults.

Velocity and Place

One of the more technical sub-fields of the mortgage and securitization industry is devoted to modelling the 'speed' of loan repayment, pre-payment, delinquency, and default. Yet such indicators are rarely studied at the local scale, or analysed spatially. Our database enables a spatially and temporally detailed view. Defaults in Essex County accelerated as the lending boom gathered momentum (Table 4.1). In 2004, the average foreclosure start was filed on a loan made almost four years earlier; this figure was almost halved by 2007, before edging up slightly.

Table 4.1 Default speeds in Essex County, New Jersey

| | | Days from origination to complaint | | |
	Number	Mean	Standard deviation	Median
2004	2,173	1,449	1,492	913
2006	2,807	941	1,223	491
2007	4,362	759	959	481
2008	5,689	880	834	691

Note: Calculations exclude un-matched geocode records and a small number of records where complaints were filed on loans originated the same day.

Source: Essex County Courthouse (2004–2008).

To explore if there is anything explicitly spatial about this accelerating crisis, we used a local indicators of spatial association (LISA) approach (see Anselin, 1995; Anselin et al., 2006), to compare the days from origination to default for each foreclosure start to the comparable times for defaults nearby. We define 'nearby' using a spatial weights matrix that includes the 25 nearest neighbours for each point. LISA calculations are used to identify specific foreclosure starts where the loan duration is correlated with the durations of nearby loans, and 999 randomized permutations across the spatial structure are used to assess whether the observed local association differs significantly (at $P = 0.05$) from what would be expected on the basis of random spatial variation. The null hypothesis suggested by options theory and risk-based pricing is that foreclosure speeds are dictated by interest rates and property values with no fundamental role for place. If we look at any one particular foreclosure start, then, the nearby cases of default should exhibit a distribution no different from the overall, countywide pattern of short, medium, and long time periods from origination to default.

LISA indices were calculated separately for 2004, 2006, 2007, and 2008, and the results are shown on the five panels of Figure 4.1 on pages 64 and 65. For the vast majority of defaults, speeds have no significant correlation with nearby defaults, indicating no significant deviation from a random spatial distribution. But some places endure localized effects. These effects intensified as the subprime boom accelerated. In 2004, one in seven foreclosure starts had a time to default that correlated with those nearby; one in four did in 2006 and 2007. By 2008 the absolute number of defaults with significant spatial effects increased over 2007 levels (11 per cent), but the increase in all defaults (30 per cent) overwhelmed the underlying spatial relations. The fine-grained patterns of default captured in Figure 4.1 provide a vivid snapshot of the local consequences of transnational capital flows (Gotham, 2009).

The most common neighbourhood effect involves faster-than-average defaults surrounded by other quick defaults. This localization occurs for about one in twenty defaults in 2004, and about twice this share in 2006 and 2007. The geography highlights predominantly African American and Latino neighbourhoods in Newark and its surrounding older, first-ring suburbs. Notable concentrations of 'quick contagion' defaults appear across a crescent of Newark neighbourhoods. The arc stretches from Lower Clinton Hill at the southern edge of the city, through West Side Park, over to the west in the Vailsburg section, to the north across Fairmont and into the more racially mixed Latino, Black, and White neighbourhoods of Roseville and Lower Broadway. A few other notable concentrations of quick-default clusters appear in the South Ironbound, the site of recent new 'Bayonne Box' style housing development and in a few parts of the inner suburbs of Irvington, Maplewood, and the City of Orange.

On the one hand, the map of quick contagion defaults highlights the contrast between the county's old industrial core in the southeast, especially Newark, and the wealthy suburban executive enclaves to the west. There are comparatively few defaults in the western suburbs (even when adjusting for the lower housing densities) and no indication of systematic spatial correlations. But the aging suburbs across the middle and eastern section of the county tell another story of divided, localized outcomes. Several distinct and statistically significant clusters appear in these older suburbs: quick defaults surrounded by loans with longer than average duration, and 'long survivor' clusters – defaults on loans originated a long time ago, surrounded by similar earlier-vintage loans.

What can we learn from this spatial analysis? In the vast majority of cases, the speed of a foreclosure shows no significant relationship to the speed of neighbouring defaults. But for about one quarter in the peak years, there are systematic neighbourhood effects. Spatial correlations highlight two kinds of neighbourhoods: a) quick-contagion defaults, mostly in working-class and lower-middle-class neighbourhoods in Newark; and b) mixtures of long-survivor defaults in older, inner-ring suburbs. These results suggest geographical processes confirming the broad outlines of other studies (Ong and Pfeiffer, 2008; Immergluck, 2009b), while revealing fine-grained variations within the categories of 'inner city' and 'inner suburb'. It is possible, however, that these patterns reflect the spatial distribution of certain types of risky borrowers, or of borrowers vulnerable to job losses or other causes of delinquency. Given the limited information in the foreclosure complaints, how can we determine whether these spatial effects persist after controlling for demand-side factors?

Modelling Local Markets

In the absence of proprietary industry datasets on long-term loan performance, the most common way to measure how lending practices affect foreclosure risk involves geographical inference. Typically, one source of data permits estimation of foreclosure rates by geographic area, while another data source measures the characteristics of the areas and/or their credit markets. We adapt this approach to focus on the role of spatial effects. We identified two sets of census tracts where foreclosure starts were dominated by statistically significant spatial effects: in one set of 19 tracts, rapid-default clusters – the rapid defaults surrounded by rapid defaults in Figure 4.1 – corresponded reasonably well to tract boundaries; in another set of nine tracts, boundaries correspond with areas dominated by mixtures of defaults surrounded by long survivors both rapid and slower. Loan-level data for the mortgage market in Essex County can then be used to answer two questions: 1) are the credit markets in areas with neighbourhood effects significantly different, after controlling for income and borrower risk profiles? 2) what credit market characteristics distinguish quick-default neighbourhoods from communities with long-survivor defaults?

Our key outcome measure is whether the loan's interest rate and up-front fees exceed HMDA's high-cost trigger, 3 percentage points above the yield on Treasury securities of comparable maturity for first-lien mortgages, and 5 points for subordinate liens. We model the likelihood that an origination will be high-cost as a function of applicant, loan, and lender characteristics using HMDA (FFIEC, 2005–2008). In addition to applicant risk measures (income, owner occupancy, loan-to-income ratio, loan purpose, etc.) we include two sets of more specialized indicators. First, we use lenders' and subsidiaries' characteristics, especially lenders' decisions on whether to sell a loan into the secondary market, to measure how local market outcomes became integrated with national and transnational capital flows. Second, we use a technique developed by Abariotes et al. (1993 see also Myers and Chan, 1995) to control for an estimate of applicant creditworthiness. The HMDA codes lenders

(a)

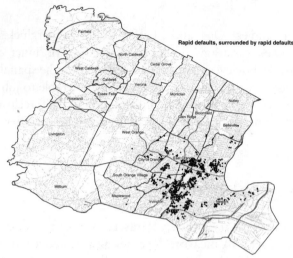

(b)

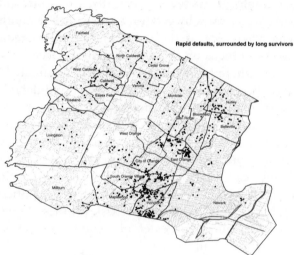

(c)

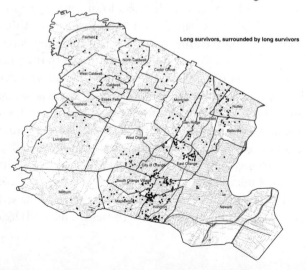

(d)
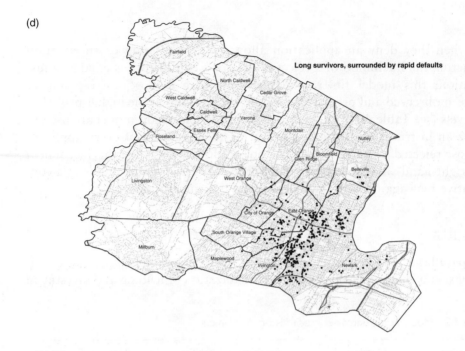

(e)
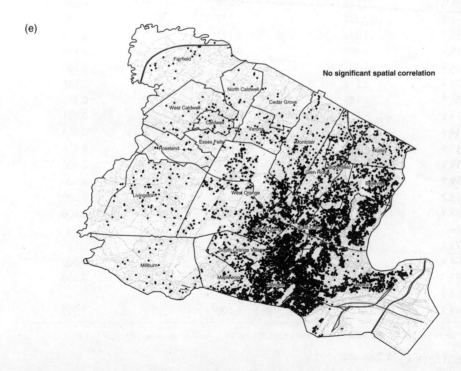

Figure 4.1 LISA indices for 2004, 2006, 2007 and 2008

use when they deny an application allow the construction of an equation predicting the likelihood of a rejection where credit history is cited as a justification; this model fits reasonably well, with a close correspondence between observed and predicted rejection rates except at the higher probability levels (see Table 4.2). The parameters of the bad-credit model are used to create an instrumental variable, measuring the similarity of every applicant to those rejected by underwriters as poor credit risks. Adding this instrument as a right-hand-side predictor in a subprime selection model builds in a conservative bias against any finding of racial discrimination.

Results

The parallax capital thesis predicts a shift from demand to supply-side explanations, and changes the geographical organization of lending and investment

Table 4.2 Model fit for borrower credit history instrument

Probability range (%)	Number of applicants	Predicted probability of bad-credit denial	Observed proportion denied for bad credit
0.1–4.9	23,876	0.030	0.031
5.0–9.9	22,235	0.072	0.070
10.0–14.9	11,211	0.122	0.112
15.0–19.9	5,489	0.173	0.174
20.0–24.9	3,453	0.223	0.260
25.0–29.9	2,517	0.274	0.286
30.0–34.9	1,181	0.320	0.310
35.0–39.9	616	0.373	0.391
40.0–44.9	398	0.422	0.354
45.0–49.9	183	0.471	0.426
50.0–54.9	35	0.521	0.400
55.0–59.9	10	0.580	0.400
60.0–64.9	4	0.626	0.500
65.0–69.9	1	0.681	1.000
70.0–74.9	–		
75.0–79.9	–		
80.0–84.9	1	0.805	0.000
85.0–89.9	1	0.855	0.000
90.0–94.9	–		
95.0–99.9	1	1.000	1.000

Notes: Max re-scaled R-squared: 0.139; per cent concordant: 73.6; number of observations: 71,212

Data source: FFIEC (2005, 2006, 2007)

networks. Model results provide support for this interpretation. Three findings stand out. First, the social inequalities of subprime lending persist after accounting for borrower characteristics. Adding the credit history instrument to the models reduces the magnitude of racial-ethnic disparities, as risk-based pricing advocates would claim. But the credit proxy, which may itself embody racially discriminatory practices and selective reporting by financial institutions, does not explain all market outcomes (see Table 4.3, Model 2). Considering credit risk and all other applicant characteristics, African Americans were 1.7 times more likely than otherwise identical non-Hispanic Whites to have high-cost loans in 2004; this disparity jumped to 2.4 in 2005, and to 2.5 in 2006. For Latinas and Latinos, the corresponding ratios went from 1.4 to 1.8 to 1.9. These inequalities are lower-bound estimates; high-cost loans are much more likely, all else constant, to have missing information on applicant race, ethnicity, or gender, and to have various kinds of data quality or editing problems (note the odds ratios between 4.3 and 6.0 for edit failures in Model 2) (Chinloy and Macdonald, 2005; cf. Immergluck, 2009a: 78–79).

Second, secondary sales networks differ between prime and subprime markets (see Table 4.3, Model 3). The GSEs rarely purchased high-cost loans (note the odds ratios between 0.10 and 0.19). Finance companies, private investors, and customized Special Purpose Vehicles (SPVs) were the key instruments in the great housing boom (note the odds ratio for SPV increased from 1.31 in 2004 to 4.51 in 2006). Wide-ranging circuits of investment and securitization constitute one aspect of parallax capital; the other entails a persistent localization of inequality. The third finding provides support for this interpretation (see Table 4.3, Model 4). The distinctive and statistically significant spatial clusters of foreclosure starts *take place* in unique local markets. Most observers recognize the importance of local variations, but view the spatial concentration of defaults and subprime loans as the concentration of risky borrowers or the culmination of abandonment by prime banks and thrifts, with the vacuum filled by mortgage companies and subsidiaries.

By including measures describing applicants and lenders, our model permits a test of this compositional effect. For the long-survivor d⸺ aging suburbs outside Newark, the local credit n by borrower and lender characteristics: all else cor are only slightly more likely to be subprime (odds 1.35) compared with the same kinds of consumers the county. Localization effects are stronger for tl Even after accounting for measures of consumer de try decisions, loans in quick-default neighbourho likely to be subprime compared with otherwise id

Countywide subprime segmentation models

Odds ratios from logistic regression

	1. Base model			2. Add credit history			3. Add securitization			4. Add spatial clusters		
	2004	2005	2006	2004	2005	2006	2004	2005	2006	2004	2005	2006
Income	0.557	0.629	0.695	0.708	0.723	0.802	0.705	0.738	0.847	0.715	0.748	0.858
Income squared	1.695	1.235	1.222	1.359	1.119	1.100	1.365	1.097	1.052*	1.347	1.090	1.045
Loan to income ratio	1.109	1.082	0.962*	1.047	1.048	0.917	1.047	1.067	0.929	1.047	1.066	0.931
Owner occupied	0.475	0.573	0.472	0.462	0.585	0.458	0.479	0.593	0.430	0.495	0.621	0.444
Subordinate lien	1.896	1.395	1.325	1.770	1.290	1.273	1.421	1.100	1.105	1.438	1.103	1.105
FHA-insured	0.089	0.066	0.058	0.078	0.054	0.056	0.089	0.051	0.050	0.088	0.051	0.050
Loan exceeds jumbo limit	0.774	0.817	1.041*	0.667	0.759	0.964*	0.566	0.638	0.810	0.577	0.655	0.828
Pre-approval requested	0.561	0.913*	0.712	0.754	1.110*	0.894*	0.712	0.927*	1.130*	0.717	0.919*	1.134
Data validity or quality edit failure	4.685	3.706	4.140	6.030	4.327	4.794	4.915	4.022	3.883	4.902	4.017	3.873
Home improvement	0.822	0.501	0.468	0.325	0.273	0.227	0.315	0.273	0.251	0.319	0.281	0.258
Refinance	1.018*	0.894	0.782	0.757	0.749	0.637	0.740	0.682	0.575	0.747	0.693	0.588
Demographic information missing	1.933	2.029	1.851	1.583	1.797	1.631	1.615	1.626	1.403	1.598	1.618	1.387
Female primary applicant	1.099	1.188	1.087	1.056*	1.154	1.058	1.056*	1.150	1.060	1.056*	1.150	1.064
Hispanic	1.828	2.229	2.367	1.361	1.854	1.937	1.317	1.646	1.652	1.302	1.629	1.623
Native American	2.511	1.699*	2.657	2.027	1.470*	2.307	1.883	1.238*	2.153	1.855	1.240*	2.152
Asian	1.025*	0.862*	0.811	0.996*	0.838	0.798	1.023*	0.830	0.815	1.027*	0.850*	0.832
African American	2.886	3.377	3.664	1.716	2.421	2.519	1.650	2.068	2.022	1.616	2.012	1.974
Lender regulated by OCC	0.189	0.239	0.217	0.135	0.196	0.188	0.117	0.217	0.210	0.116	0.217	0.211
Lender regulated by OTS	0.190	0.357	0.484	0.159	0.340	0.459	0.135	0.327	0.484	0.135	0.327	0.485
Lender regulated by FDIC	1.808	1.505	0.889*	1.675	1.515	0.933*	1.081*	0.875*	0.633	1.079*	0.876*	0.632
Lender reports to HUD	2.489	2.998	1.564	2.469	3.059	1.625	1.623	2.221	1.011*	1.611	2.204	1.002
Credit history instrument				1.624	1.376	1.425	1.557	1.461	1.572	1.546	1.451	1.567

	Odds ratios from logistic regression											
	1. Base model			2. Add credit history			3. Add securitization			4. Add spatial clusters		
	2004	2005	2006	2004	2005	2006	2004	2005	2006	2004	2005	2006
Loan sold to affiliate institution							0.719	1.262	1.693	0.720	1.261	1.694
Loan sold to bank							0.952*	2.243	1.033*	0.951*	2.238	1.040
Loan sold to finance company							0.629	1.646	3.459	0.631	1.665	3.475
Loan sold to GSE							0.104	0.163	0.189	0.104	0.164	0.192
Loan sold to other purchaser (SPV)							1.312	2.256	4.513	1.335	2.298	4.470
Loan sold to private investor							1.002*	3.515	4.812	0.995*	3.500	4.828
Rapid-default census tract cluster										1.889	2.187	1.839
Long survivor census tract cluster										1.280	1.354	1.177
Rapid default * Loan sold to SPV										0.683	0.830*	1.210
Long survivor * Loan sold to SPV										0.984*	0.906*	1.009
Max-rescaled Pseudo-R-squared	0.31	0.34	0.29	0.32	0.35	0.30	0.37	0.42	0.41	0.37	0.42	0.41

Number of observations:
2004: 24,807 (20.4% high cost)
2005: 32,181 (37.1%)
2006: 30,973 (40.8%)

Notes:
For continuous variables, figures report standardized odds ratios (change in odds with an increase in the predictor variable by one standard deviation)
Reference categories: home purchase (purpose), non-Hispanic White (race/ethnicity), national bank regulated by Federal Reserve (regulator), loan not sold in same calendar year as origination (loan sales)
*P > Wald Chi-square > 0.05; all other odds ratios are based on coefficients significant at P<0.05

Data source: FFIEC (2005, 2006, 2007)

This two-to-one disparity reminds us that the transnational facets of investment that fuelled the subprime boom were woven into communities: even as loans were packaged into securities for sale to global investors, they relied on local lending infrastructures, borrowers, and places.

The spatial localization effects that highlight a two-to-one disparity for quick-default clusters persist after controlling for lender type and secondary sales circuits. Even so, adding an interaction term reveals the interplay of local and global processes. At the height of the boom, when loans sold to Wall Street SPVs were four and a half times more likely to be subprime across the entire county, the SPV sales in quick-default clusters were even *more* likely to be subprime (see the 1.21 odds ratio near the bottom of Table 4.3, Model 4). Some interpretations of the financial crisis suggest that some originators operated an originate-to-sell model with less incentive to scrutinize loan quality. The effect of clustering so many risky loans in some places is evident in the landscape of boarded up homes in these communities.

The short period from 2004 to 2006 brought an intensification of the racial and ethnic inequalities that have long been among the most local of American social relations. At the same time, the links tightened between local outcomes and aggressive global banking and investment decisions as changes within the financial institutional structure made it possible to connect global capital circuits to urbanization processes directly, with great speed, and little regulation.

These realignments sharpened contrasts between wealthy suburbs and distressed urban neighbourhoods (Brown et al., 2009; Crump et al., 2008) and the vast metropolitan regions of Florida and the Southwest (Wyly et al., 2009). And they etched divisions *among* and *within* formerly disinvested urban communities. To explore these divisions, we estimated loan-level models of differences between subprime borrowers in quick-default clusters versus those in long-survivor neighbourhoods. To capture the localized social geographies, we combined HMDA codes on race, ethnicity, gender, and presence/absence of co-applicant. Stepwise logistic variable selection was used to identify statistically significant indicators that distinguish subprime borrowers in quick-default neighbourhoods from their subprime peers in long-survivor communities (Table 4.4). Borrowers in quick-default clusters were much less likely to be owner occupiers (0.29) or to seek loans for renovation (0.39) or refinance (0.43). In general, the patchworks of concentrated quick defaults shown in Figure 4.1 highlight areas where moderate- and middle-income borrowers used subprime credit for home purchase: subprime borrowers in these areas had a median annual income of $85,000 (2005 metropolitan area median family income was $80,300), and 75 percent obtained loans on properties they did not intend to occupy. Borrowers in the long-survivor neighbourhoods reported median annual incomes of $77,000, and 90 per cent of them were owner-occupiers; 60 per cent were existing owners, taking out loans for home improvement or refinance.

Table 4.4 Contrasts between quick-default and long-survivor clusters

	Odds ratio predicting likelihood of loan in a quick-default cluster	Mean values	
		Quick-default clusters	Long-survivor clusters
Loan to income ratio	0.90	0.882	0.876
Owner occupied	0.29	0.753	0.907
Loan exceeds jumbo limit	0.46	0.019	0.041
Home improvement	0.39	0.037	0.076
Refinance	0.43	0.341	0.541
Non-Hispanic White Couple, Male primary applicant	0.48	0.007	0.012
Non-Hispanic White Couple, Female primary applicant	0.26	0.001	0.006
Non-Hispanic White Male	1.73	0.108	0.044
Non-Hispanic Black Couple, Male primary applicant	0.29	0.024	0.069
Non-Hispanic Black Couple, Female primary applicant	0.28	0.012	0.040
Non-Hispanic Black Male	0.68	0.204	0.216
Non-Hispanic Black Female	0.49	0.149	0.239
Non-Hispanic Black, gender unreported and other	0.34	0.013	0.032
Hispanic White Couple, Male primary applicant	3.36	0.014	0.004
Hispanic White Couple, Female primary applicant	6.16	0.004	0.0009
Hispanic White Male	3.92	0.120	0.024
Hispanic White Female	2.18	0.058	0.023
Hispanic Black Female	0.33	0.003	0.006
Credit history instrument	1.12	0.077	0.100
OTS regulated thrift	0.62	0.036	0.057
Lender national market share	0.90	0.977	1.240
Lender share Hispanic	1.09	0.211	0.189
Lender share of originations sold to other purchaser (SPV)	1.08	0.316	0.265
Loan sold to GSE	0.55	0.010	0.018
Max-rescaled Pseudo-R-squared	0.23		
Number of observations:	4,573		

Notes: For continuous variables, figures report standardized odds ratios (change in odds with an increase in the predictor variable by one standard deviation). Reference categories: home purchase (purpose), unidentified (race/ethnicity), independent mortgage company (regulator); loan not sold in same calendar year as origination (loan sales) Variables identified with stepwise selection. All odds ratios are based on coefficients significant at $P < 0.05$

Data source: FFIEC (2005, 2006, 2007)

The class composition of high-risk lending is interwoven with race, ethnicity, gender, and family circumstances. The earlier findings (Table 4.3, Models 3 and 4) confirm that lending industry structure and secondary market circuits of capital investment are decisive in explaining the subprime market. When the view is shifted away from the structured exploitations of the prime/subprime divide, and refocused on differences *among subprime borrowers*, a different picture becomes clear. Circuits of lending, securitization, and investment were built atop local geographies. The most decisive contrasts between borrowers in quick-default and long-survivor clusters involve relations of race, ethnicity, gender, and presence of co-applicants. The models distinguish a mixed White–Latino market of homebuyers and small-scale investors in the quick-default clusters, and a community of non-Hispanic African American homeowners drawn into home equity, renovation, and refinance loans in the crescent of Newark and inner-suburban neighbourhoods. Inequalities of race and class are exacerbated by inequalities of gender: in a separate detailed countywide model, a statistically significant interaction term indicates that female primary applicants are 1.5 times more likely in the long-survivor default cluster neighbourhoods to wind up with a subprime loan compared with otherwise identical female applicants elsewhere across the county.

Conclusions

Early revelations of systemic mortgage problems emerged in early 2007, and eventually culminated in a global credit crisis. Nation-states and multilateral organizations sought to stabilize financial institutions, resuscitate credit flows, and relieve a liquidity crisis. The imperative to avert global economic crisis ignored the urbanization processes and the consequent locally experienced outcomes. Substantial 'investments' to bail out financial institutions dwarfed the assistance to borrowers, renters, and communities.

For two decades, the return of capital to the city was celebrated as a sign of policy success and market benevolence. Cities were back! But in many places, those gains were illusory. Looking back, it is easier to make out the connections between the broad relationship between the global financialized economy and urbanization. An urban politics of finance can help to uncover the political economy of urban change as it is tied to global capital flows and untangle the policy decisions that enable capital to reach communities in the way that it does. At the moment, the problems within communities look local, fragmented, and caused by the people in these places. An urban politics of finance can tie these local experiences together to systematically understand the policy choices made that constructed a political economy in which the foreclosure crisis was possible.

While efforts to enhance disclosure during the mortgage process and financial education are welcomed by many, efforts on this front do not address the political economy that made the crisis possible in the first place. By illuminating the connection between capital and community, an urban politics of finance can not only better articulate the sources of change in the post-industrial city, they can create a space where it is possible to consider a response.

References

Abariotes, A., Ahuja, S. Feldman, H. Johnson, C. Subaiya, L. Tiller, N. Urban, J. and Myers, S.L. Jr. (1993) Disparities in Mortgage Lending in the Upper Midwest. Paper presented at the Fannie Mae University Colloquium on Race, Poverty, and Housing Policy. Minneapolis, MN, December 3.

Ambrose, B.W. and Capone, C.A. (1996) 'Cost–Benefit Analysis of Single-Family Foreclosure Alternatives', *Journal of Real Estate Finance and Economics*, 13(2): 105–119.

Ambrose, B.W., Buttimer, R.J. and Capone, C.A. (1997) 'Pricing Mortgage Default and Foreclosure Delay', *Journal of Money, Credit, and Banking*, 29(3): 314–325.

Anselin, L. (1995) 'Local Indicators of Spatial Association – LISA', *Geographical Analysis* 27(2): 93–115.

Anselin, L., Syabri, I. and Kho, Y. (2006) 'GeoDa: An Introduction to Spatial Data Analysis', *Geographical Analysis* 38(1): 5–22.

Apgar, William, Calder, Allegra and Fauth, Gary (2004) *Credit, Capital, and Communities: The Implications of the Changing Mortgage Banking Industry for Community Based Organizations*. Harvard Joint Center for Housing Studies: Cambridge, MA.

Ashton, P. (2009) 'An Appetite for Yield: The Anatomy of the Subprime Mortgage Crisis', *Environment and Planning A* 41(6): 1420–1441.

Avery, R., Bostic, R., Calem, P. S., Canner, G.B. (1996) 'Credit Risk, Credit Scoring, and the Performance of Home Mortgages', *Federal Reserve Bulletin*, July: 621–648.

Brown, Larry, Webb, Michael and Chung, Su-Yeul (2009) 'Population Vulnerability in a Contemporary Setting: Housing Foreclosure in Columbus, Ohio, 2003–2007', paper presented at the annual meeting of the Association of American Geographers, Las Vegas, NV, April.

Bunce, Harold L., Gruenstein, Debbie, Herbert, Christopher E. and Scheessele, Randall E. (2001) *Subprime Foreclosures: The Smoking Gun of Predatory Lending?* U.S. Department of Housing and Urban Development: Washington, DC.

Chinloy, P. and Macdonald, N. (2005) 'Subprime Lenders and Mortgage Market Completion', *Journal of Real Estate Finance and Economics*, 30(2): 153–165.

Cox, K.R. (1993) 'The Local and the Global in the New Urban Politics: A Critical View', *Environment and Planning D*, 11(4): 432–448.

Crump, J., Newman, K., Belsky, E.S., Ashton, P., Kaplan, DH., Hammel, DJ. and Wyly, E. (2008) 'Cities Destroyed (Again) for Cash: Forum on the U.S. Foreclosure Crisis', *Urban Geography* 29(8): 745–784.

Dahl, Robert (1961) *Who Governs?* Yale University Press: New Haven, CT.

Danis, M.A. and Pennington-Cross, A. (2008) 'The Delinquency of Subprime Mortgages', *Journal of Economics and Business*, 60(1–2): 67–90.

DeFilippis, J. (1999) 'Alternatives to the 'New Urban Politics': Finding Locality and Autonomy in Local Economic Development', *Political Geography* 18(8): 973–990.

Deng, Y., Quigley, J., Van Order, R. (1996) 'Mortgage Default and Low Downpayment Loans: The Costs of Public Subsidy', *Regional Science and Urban Economics* 26 (3–4): 263–285.

Duncan, Douglas (2007) 'MBA Testifies Before Senate: Market is Working, Benefiting Consumers and Increasing Homeownership', press release, 7 February. Mortgage Bankers Association of America: Washington, DC.

Dymski, G.A. (1999) *The Bank Merger Wave*. M.E. Sharpe: Armonk, NY.

Dymski, G.A. (2009) 'Mortgage Markets and the Urban Problematic in the Global Transition', *International Journal of Urban and Regional Research* 33(2): 427–442.

Elmer, P. and Seelig, S. (1999) 'Insolvency, Trigger Events, and Consumer Risk Posture in the Theory of Single-Family Mortgage Default', *FDIC Working Paper* 98–3.

Engel, K.C. and McCoy, P.A. (2002) 'A Tale of Three Markets: The Law and Economics of Predatory Lending', *Texas Law Review* 80: 1255–1381.

Engel, K.C. and McCoy, P.A. (2007) *The Legal Infrastructure of Subprime and Non-Traditional Home Mortgages*. Harvard Joint Center for Housing Studies: Cambridge, MA.

Federal Financial Institutions Examination Council (FFIEC) (Annual) Home Mortgage Disclosure Act, Loan Application Register Raw Data on CD-ROM. FFIEC: Washington, DC.

Foote, Christopher L., Gerardi, Kristopher and Willen, Paul S. (2008) 'Negative Equity and Foreclosure: Theory and Evidence', Federal Reserve Bank of Boston Public Policy Discussion Paper 08-3.

Foster, C. and Van Order, R. (1985) 'FHA Terminations: A Prelude to Rational Mortgage Pricing', *AREUEA Journal* 13(3): 273–291.

Gotham, K.F. (2009) 'Creating Liquidity Out of Spatial Fixity: The Secondary Circuit of Capital and the Subprime Mortgage Crisis', *International Journal of Urban and Regional Research* 33(2): 355–371.

Immergluck, D. (2004) *Credit to the Community*. M.E. Sharpe: Armonk, NY.
Immergluck, D. (2009a) *Foreclosed: High-Risk Lending, Deregulation, and the Undermining of America's Mortgage Market*. Cornell University Press: Ithaca, NY.
Immergluck, D. (2009b) 'The Foreclosure Crisis, Foreclosed Properties, and Federal Policy', *Journal of the American Planning Association*, 75(4): 1–18.
Immergluck, Dan and Smith, Geoff (2004) *Risky Business: An Econometric Analysis of the Relationship Between Subprime Lending and Neighborhood Foreclosure*. Woodstock Institute: Chicago.
Jonas, Andrew E.G. and Wilson, David (eds) (1999) *The Urban Growth Machine: Critical Perspectives Two Decades Later*. State University of New York Press: Albany, NY.
Levitin, Adam and Wachter, Susan (2010) 'Explaining the Housing Bubble', Working Paper downloaded from Social Science Research Network Electronic Paper Collection at: http://papers.ssrn.com/sol3/papers.cfm?abstract_id=1669401.
Levitin, A. and Wachter, S. (2011) 'Information Failture and the U.S. Mortgage Crisis' in S. Wachter and M. Smith, *The American Mortgage System: Crisis and Reform*. University of Pennsylvania Press: Philadelphia.
Martin, D. (2011) 'Urban Politics as Sociospatial Struggles', *International Journal of Urban and Regional Research* 35(4): 853–871.
Mian, A. and Sufi, A. (2009) 'The Consequences of Mortgage Credit Expansion: Evidence From the U.S. Mortgage Default Crisis', *Quarterly Journal of Economics*, November: 1449–1496.
Myers Jr., S.L. and Chan, T. (1995) 'Racial Discrimination in Housing Markets: Accounting for Credit Risk', *Social Science Quarterly* 76(3): 543–561.
Ong, P. and Pfeiffer, D. (2008) *Spatial Variation in Foreclosures in Los Angeles*. Department of Urban Planning, University of California: Los Angeles.
Orderud, G.I. (2011) 'Finance, Home Building, and Urban Residential Structuring', *Antipode* 43(4): 1215–1249.
Quigley, J.M. and Van Order, R. (1995) 'Explicit Tests of Contingent Claims Models of Mortgage Default', *Journal of Real Estate Finance and Economics*, 11: 99–117.
Schwartz, L. (2005) 'Roth, Race, and Newark', *Cultural Logic*, 8.
Squires, Gregory D. (ed.) (2003) *Organizing Access to Capital*. Temple University Press: Philadelphia, PA.
Stone, C.N. (1993) 'Urban Regimes and the Capacity to Govern: A Political Economy Approach', *Journal of Urban Affairs* 15(1): 1–28.
Straka, J.W. (2000) 'A Shift in the Mortgage Landscape: The 1990s Move to Automated Credit Evaluations', *Journal of Housing Research* 11(2): 207–232.
U.S. GAO (2007) 'Home Mortgage Defaults and Foreclosures: Recent Trends and Associated Economic and Market Developments', Briefing to the Committee on Financial Services. Housing of Representatives. 10 October.

Vandell, K.D. (1993) 'Handing Over the Keys: A Perspective on Mortgage Default Research', *Journal of the American Real Estate and Urban Economics Association* 21(3): 211–246.

Wachter, S. (2009) 'The Ongoing Financial Upheaval: Understanding the Sources and Way Out', University of Pennsylvania Law School. Institute for Law and Economics. Research Paper No. 09-30.

Wyly, E., Moos, M., Hammel, D. and Kabahizi, E. (2009) 'Cartographies of Race and Class: Mapping the Class-Monopoly Rents of American Subprime Mortgage Capital', *International Journal of Urban and Regional Research*, 33(2): 332–354.

SECTION 2
CITY AS MEDIUM

INTRODUCTION TO CITY AS MEDIUM

Mark Davidson and Deborah Martin

medium [me-di-um] ~ an intervening substance through which a force acts or an effect is produced

This section of the book focuses on the idea that the city functions as a medium through which politics are enacted. As such it identifies the ways in which actors and groups of actors utilize the city as a means to achieve certain ends. Our bracketing of 'urban politics' here is therefore less geographical than the previous section. This is not to say that we are not interested in the geography of the city. But rather we first look at the actors who utilize the city as a particular environment. In the section you will read about the ways in which mayors, political institutions and legal systems use the city as a means to achieve their ends.

There are many strands of the urban politics literature that look at how various actors are empowered and mobilized through their intervention in the urban process. Perhaps the most pronounced strand of thought in this genre is urban regime theory. This theory of urban politics focuses on explaining how and why certain actors come to dominate the shape and form of city government. This understanding of urban politics is often credited to Clarence Stone's 1989 study of Atlanta, Georgia. Stone sees urban regime theory as based within political economy perspectives in that it is concerned with the economic forces that make and implement distinct policy initiatives. The theory can therefore be said to have two elements: a concern with the external economic factors (i.e. market competition, competitive bidding for national government resources) and internal political dynamics (i.e. coalition building between parties). Urban regime theory is therefore concerned with those occurrences where various actors come together within institution settings to achieve a particular goal. This goal might be the construction of a particular retail space, the building of a new road or the blocking of certain environmental regulations. In each case, we need to be cognizant of the ways in which certain actors come together to shape these urban political dynamics.

The urban scholarship that has taken up and developed urban regime analysis has focused upon actors operating within and upon urban government. Often times this is concerned with the connections between political and economic actors with particular cities, whereby those with vested interests in economic development seek to shape and direct the actions of democratically elected governments (e.g. Logan and Molotch, 1987; Hackworth, 2002; Jonas and McCarthy, 2009). But there is another strand of urban scholarship that looks at the ways in which certain individuals and/or groups use the fabric of the city to achieve certain goals. These would include those who have examined how the processes of building and regulating urban environments are bound up with processes of social regulation and control.

This section of the book contains three different examples of the city being used as a medium for certain political actors. In Chapter 5, Kurt Iveson takes up the notion that the city is a policed space. For those familiar with the past two decades of urban scholarship, this idea of the city as a policed space will bring to mind the theoretical influence of French philosopher and social theorist Michel Foucault. Throughout the 1990s and 2000s, the work of Foucault greatly influenced the social sciences. Foucault's studies of the birth of modern society revealed the numerous ways in which social order and regulation was constructed and re-created. For example, his work on mental illness and prisons illuminated the mechanisms of social ordering and regulation that were generated alongside the birth of the industrial city. From this body of work, Foucault's deconstruction of Jeremy Bentham's eighteenth-century prison design, the panopticon, has proven particularly influential:

> For Foucault the Panopticon represented a key spatial figure in the modern project and also a key *dispositive* in the creation of modern subjectivity, in other words in the remaking of people (and society) in the image of modernity. Panopticism, the social trajectory represented by the figure of the Panopticon, the drive to self-monitoring through the belief that one is under constant scrutiny, thus becomes both a driving force and a key symbol of the modernist project.
> (Wood, 2003: 235)

The use of this prison model as a metaphor for the ways in which urban space has been constructed in order to generate certain behaviours and social norms provides an important entry point for thinking about the politics of urban space.

BOX S2 FOUCAULT AND THE MEDIUMS OF SOCIAL CONTROL

Michel Foucault (1926–1984) was a French philosopher and social theorist who was particularly interested in the history of knowledge, social institutions and modernity. He is regarded by many as one of the most influential thinkers of the twentieth century. His first work, *Madness and Civilization*, examined how Western societies thought about and dealt with human madness. By tracing out this history, Foucault demonstrated how certain ideas had been pivotal in constructing an understanding of what madness is. This method of social anthropology, or 'genealogy' as he termed it, was used by Foucault in later studies to understand the emergence and societal role of medicine, the sciences, punishment and sexuality. A fundamental argument he made was that the concepts we know through language shape our very understandings of our social world, including understandings of ourselves as 'subjects' who think and know. As the Anglo-American social sciences began to engage with post-structuralist theory in the 1980s, Foucault became an important theoretical influence in disciplines such as geography, cultural studies and sociology. His work remains important today for critical scholars interested in examining how power operates, and how language and practice – discourses – shape knowledge itself. There is many urban scholars who continue to draw on Foucault's theories; below is listed just a small selection of representative writings you might want to explore:

Urban politics texts

Conradson, D. (2003) 'Spaces of care in the city: the place of a community drop-in centre', *Social and Cultural Geography*, 4(4): 507–525.
Gieryn, T. (2002) 'What buildings do', *Theory and Society*, 31(1): 35–74.
Pløger, J. (2008) 'Foucault's dispositif and the city', *Planning Theory*, 7(1): 51–70.

General audience texts

Crampton, J. and Elden, S. (eds) (2007) *Space, Knowledge and Power: Foucault and Geography*. Ashgate: Aldershot.
Philo, C. (1992) 'Foucault's geography', *Environment and Planning D: Society and Space,* 10(2): 137–161.

These politics are developed in Iveson's chapter through a discussion of graffiti and related attempts to regulate the activity. Iveson bases his discussion of graffiti within Jacques Rancière's theory of politics. We've already seen Rancière employed by Hankins and Martin to discuss the politics of strategic neighbouring. But here Iveson focuses on what Rancière calls the opposite of politics: policing. Rancière's usage of the term police/policing has much in common with Foucault's work on social regulation in modern society. For Rancière the police are all those things which go into maintaining a particular social order: rules, laws, social norms, acceptable behaviours, roles, identities, and so on. And Foucault was interested in the ways in which things such as medical knowledge, prisons and sexual identities all went into shaping a certain social regime. In Iveson's discussion of graffiti he explains how various mechanisms go into regulating graffiti, from explicit acts of policing to moral codes within graffiti communities. The point to be identified here, then, is that in policing the city many elements come together to maintain a social regime. It is not purely the state or institutional actors that create society and space, but also the actors themselves. If we are to engage with questions of struggle and political change, what Iveson's illustrative example demonstrates is the multitude of elements that must be considered. Or, to put it in Rancière's terms, for politics to occur the police order itself must be transcended and another, more just, one installed.

In Donald McNeill's chapter, our tack changes slightly. McNeill's chapter focuses on the mayor as an important political agent. The chapter argues that the mayor is an often under-appreciated figure in urban politics. Whereas approaches such as urban regime theory focus on coalitions of interests, McNeill argues that it can, in many cases, be the actions of one person that end up directing urban political change. A particularly enigmatic, wealthy and/or well-connected mayor can exert power within cities that far outweigh their institutional role. The argument here is not that mayors are the only figure of political power in some cities. Indeed, McNeill stresses that those everyday disciplining processes that the likes of Foucault have illustrated are usually central to understanding political power. Rather the chapter's central contribution is in demonstrating that, on occasion, the city can become a medium through which a particular individual exerts an influence that changes the city beyond that which can be explained by structural conditions. In this sense the city becomes a medium for that powerful individual to impose a vision of the city.

The final chapter of the section focuses squarely on how institutions of social regulation are enacted through the city. John Carr takes us to Seattle, Washington; to the scene of public planning battles over the provisioning of skate parks.

From his insider's perspective, Carr argues that what might be considered a struggle over collective provision actually turns out to be a lesson over how social control operates in and through legal structures and procedures. The chapter's emphasis on the role of law in the (non-)creation of urban politics is part of a broader move across urban studies and geography to understand the role of law in the urban process (e.g. Blomley, 2004; Martin et al., 2010). Within this literature legal systems are understood as normative agents that actively reproduce a certain type of society and city. In Carr's chapter, this is a society and city that do not respond to democratic decision making. Rather, the legal system is used by certain actors to maintain a status quo and protect privileged interests. As Carr urges us to recognize, we cannot see community consultation practices as neutral devices that serve democratic ends. Rather, they function to pacify citizens and stymie political change.

The chapters in this section therefore take us beyond the collection of actors we most commonly associate with urban politics. Urban politics do not serve simply as a medium through which the interests of place-based capital are collectively mobilized. Rather, the chapters show illustrative examples of the ways in which the city is used by various actors in various ways to produce certain outcomes, or in some cases to ensure certain outcomes are not produced. The chapters all do this in different ways. For McNeill, it is the individual persona of the mayor who must be recognized as a potentially powerful political agent. For Carr, it is the legal regime in Seattle that serves particular interest over others. And for Iveson, his drawing on Jacques Rancière's theory of politics and policing directs him to see societal-wide modes of social regulation imposing themselves through a set of different actors. We therefore have different types and scales of political action engaged in using the city as a medium of social power and control. We might then think of the following questions:

- If the city is used by multiple actors to achieve certain ends, who are the most powerful actors and what are their goals?
- Do the institutions of urban politics, mayors and councils, have influence over all political processes in the city?
- Is everything political in the city? Or are things rarely political in the city?
- If mayors have significant personal power and/or democratic processes of citizen engagement do not work, how can we assess the justness of our political institutions?
- How compatible are the political theories of people like Jacques Rancière with established theories of urban politics, such as urban regime theory?

References

Blomley, Nicholas (2004) *Unsettling the City: Urban Land and the Politics of Property*. Routledge: London.

Hackworth, J. (2002) 'Local autonomy, bond-rating agencies and neoliberal urbanism in the United States', *International Journal of Urban and Regional Research*, 26(4): 707–725.

Jonas, A.E.G. and McCarthy, L. (2009) 'Urban management and regeneration in the United States: state intervention or redevelopment at all costs?', *Local Government Studies*, 35(3): 299–314.

Logan, John R. and Molotch, Harvey (1987) *Urban Fortunes: The Political Economy of Place*. University of California Press: Berkeley, CA.

Martin, D.G., Scherr, A. and City, C. (2010) 'Making law, making place: lawyers and the production of space', *Progress in Human Geography*, 34(2): 175–192.

Stone, Clarence (1989) *Regime Politics: Governing Atlanta 1946–1988*. University of Kansas Press: Lawrence, KA.

Stone, C. (1993) 'Urban regimes and the capacity to govern: a political economy approach', *Journal of Urban Affairs*, 15(1): 1–28.

Wood, D. (2003) 'Foucault and panopticism revisited', *Surveillance and Society*, 1(3): 234–239.

5
POLICING THE CITY

Kurt Iveson

Introduction

> The police can procure all sorts of good, and one kind of police may be infinitely preferable to another. This does not change the nature of the police ... Whether the police is sweet and kind does not make it any less the opposite of politics. (Jacques Rancière, *Disagreement*, 1999: 31)

In this chapter, my broad aim is to consider what an account of *policing* the city might add to our thinking on urban politics. The shared etymological roots of the English words politics and police (not to mention polity and policy) in the Ancient Greek word πόλις (*polis*) provides us with a clue about their close relation, each capturing some dimension of being together in a political community. In fact, I want to argue that no discussion of urban *politics* would be complete without some consideration of *police* – to do urban politics is to attempt to undo the naturalized order of the city that is established through the policing of places and the individual and collective bodies that inhabit and produce them.

To introduce a discussion of the *relationship* between policing and politics with a quotation which asserts that police is the *opposite* of politics might seem odd. But Jacques Rancière's concept of 'the police' can be very helpful indeed in thinking about the policing of cities, and this framework for thinking about the policing of cities is also helpful in thinking about the nature and possibilities of urban politics.[1] Following Rancière, I will argue that policing is best conceptualized procedurally rather than institutionally. The policing of cities is not exclusively a matter for commissioned police officers – rather, policing refers to a wider set of procedures which seek to allocate and contain particular bodies and behaviours to their 'proper' places in the city. As such, it exists in an oppositional relationship to politics, which works to contest these allocations and containments.

The chapter will illustrate the usefulness of this approach to policing through a discussion of the policing of graffiti. This discussion will consider the variety of actors who are involved in efforts to police the placement of graffiti in cities – from uniformed police and private security guards, to planners, urban designers, architects, property owners, art collectors and gallery owners, social workers and graffiti writers themselves. The chapter will then reflect on the potential for a politics of graffiti to emerge out of these tensions and contradictions. But, first, let me elaborate on Rancière's conception of policing and its usefulness for our understanding of urban politics.

Politics and Police

For Rancière, politics and police are two opposing logics or modes of being together. 'The police', in the broad sense adopted by Rancière, refers to an order or community in which everyone has been assigned a part to play and a proper place in which to play it. According to the police, the 'whole' of the community is no more than the sum of these identified 'parts'. The police logic is the one that 'simply counts the lots of the parties, that distributes bodies within the space of their visibility or their invisibility and aligns ways of being, ways of doing, and ways of saying appropriate to each' (Rancière, 1999: 28). Politics, on the other hand, is 'the logic that disrupts this harmony' (ibid.). It exists 'wherever the count of parts and parties of society is disturbed by the inscription of a part of those who have no part' (Rancière, 1999: 123). When those who have no part in the police order insist that in fact they do have a part, they demonstrate that the police order is actually a miscount of the community – that the parts assigned by the police order do not in fact add up to the whole. Politics, then, is defined by Rancière as a particular kind of disruption or contestation of the police order:

> Politics exists because those who have no right to be counted as speaking beings make themselves of some account, setting up a community by the fact of placing in common a wrong that is nothing more than this very confrontation, the contradiction of two worlds in a single world: the world where they are and the world where they are not, the world where there is something 'between' them and those who do not acknowledge them as speaking beings who count and the world where there is nothing. (1999: 27)

Starting from this understanding of the police and politics, three important points follow. First, much of what is often given the name *politics* is actually a form of police. As Rancière puts it:

> Politics is generally seen as the set of procedures whereby the aggregation and consent of collectivities is achieved, the organization of powers, the distribution of places and roles, and the systems for legitimizing this distribution. I propose to give this system of distribution and legitimization another name. I propose to call it the police. (1999: 28)

Politics, in other words, is certainly not something done exclusively (or even at all!) by politicians, who are frequently engaged in efforts to police society by consolidating and managing the existing order, and allocating proper places to the groups they recognize in the polity.

Second, policing is not something just done exclusively by the police. In his discussions of the police, Rancière is careful to distinguish between the police order and the 'uniformed' or 'petty' police. As he puts it, 'The petty police is just a particular form of a more general order that arranges that tangible reality in which bodies are distributed in the community' (1999: 28).

Third, and perhaps most importantly, while politics and police are distinct from one another, they cannot exist without each other. As Rancière insists, 'if politics implements a logic entirely heterogenous to that of the police, it is always bound up with the latter' (1999: 31). As I have just noted, neither politics nor police describe particular groups of people; rather, they refer to different logics of being together. So, while politics is always oppositional to an extant police order, this is not a simple opposition between two sides, as if between two teams in a sporting contest. Rather, the politics/police distinction can be deployed in order to explore the ways in which these two logics interact:

> If the distinction between politics and the police can be useful, it is not to allow us to say: politics is on this side, police is on the opposite side. It is to allow us to understand the form of their intertwinement. We rarely, if ever, face a situation where we can say; this is politics in its purity. But we ceaselessly face situations where we have to discern how politics encroaches on matters of the police and the police on matters of politics. (Rancière, 2009: 287)

This understanding of the police and its relationship to politics is a useful framework for thinking about the policing of cities. The police order and politics have a fundamentally spatial dimension (Dikeç, 2005). The police order involves the allocation of people and practices to their 'proper place', such that places themselves have their proper meaning, users and uses (Dikeç, 2005: 174; see also Cresswell, 1996). Politics occurs when this allocation is confronted with another. For Rancière, 'Politics acts on the police. It acts in the places and with the words that are common to both, even if it means reshaping those places and changing the status of those words' (1999: 33). Rancière offers several illustrations of the 'reshaping of places', such as this short discussion of the street:

> The police says that there is nothing to see on a road, that there is nothing to do but move along. It asserts that the space of circulating is nothing other than the space of circulation. Politics, in contrast, consists in transforming this space of 'moving-along' into a space for the appearance of a subject: i.e., the people, the workers, the citizens: It consists in refiguring the space, of what there is to do there, what is to be seen or named therein. (2001: 22)

Elsewhere (Rancière, 2006), he considers the struggle of workers to transform the workplace from a private place where terms and conditions of work are set by owners and managers and offered to workers, to a public place where terms and conditions have to be negotiated with workers who have rights based on their common humanity. In summary, from this perspective politics has a very particular relationship to space, as space: 'becomes an integral element of the interruption of the 'natural' (or, better yet, naturalized) order of domination through the constitution of a place of encounter by those that have no part in that order' (Dikeç, 2005: 172).

I now want to demonstrate the usefulness of this perspective on the police by applying it to the case of graffiti. I will proceed from here by first placing graffiti in relation to the police order of the contemporary capitalist city. Then I will consider a variety of ways in which graffiti is 'policed', drawing on the particular sense of 'police' advanced by Rancière. Finally, the chapter will conclude with some reflections on what the disagreement over graffiti might teach us about the nature of urban police and its relationship to urban politics.

'Imagine a city ...'

> People say there is a graffiti problem. The only problem is that there isn't enough of it. Imagine a city where graffiti wasn't illegal, a city where everybody could draw wherever they liked. Where every street was awash with a million colours and little phrases. Where standing at a bus stop was never boring. A city that felt like a living breathing thing which belonged to everybody, not just the estate agents and barons of big business. Imagine a city like this and stop leaning against that wall – it's wet. (Banksy, *Existencilism*, 2000)

Urban landscapes are media landscapes. Among the many forms of mediated communication that take place in and through cities, all manner of actors put the surfaces of cities to work as spaces of written communication. As urban inhabitants move through the city, they are addressed by public and private authorities, property owners, advertisers, activists, neighbours, graffiti writers, street artists and many others who have installed words and images in a variety of formats.

This urban media landscape is caught up with efforts to police the city in two related senses. First, many instances of written communication in the city are intended to police the behaviour of urban inhabitants, in what Hermer and Hunt (1996) call a kind of 'regulation at a distance'. Second, the outdoor media landscape is itself also subject to policing, as various urban authorities attempt to define and enforce limits on the authorized use of urban surfaces as media. The policing of the urban media landscape thereby seeks to establish a fixed relationship between authorship and authority. These policing efforts are concerned with regulating who is able to write, what they are able to write and where they are able to write.

Of course, these claims to authority and the policing efforts that are enacted by authorities are frequently contested and/or ignored (see Figure 5.1). But many of these disputes are not *political* as such, in the particular sense implied by Rancière. For example, there are debates between advertisers and planning officials about the location of outdoor advertising in the city, and there are debates between particular advertisers and place-managers over the content of their advertisements. But in such debates, there are is no particular challenge to the police order, in which property ownership and planning authority define which parts of the community have access to the urban media landscape.

Figure 5.1   A sign prohibiting graffiti, covered in and surrounded by graffiti, Sydney (source: author)

I think it is fair to say that in recent years, the most significant political challenge to the authority of the authorities to police the urban media landscape has come from graffiti writers and artists. Drawing on the framework for understanding of urban politics and policing outlined earlier in this chapter, we can see graffiti writers and artists as instigating (although not always conducting) a kind of urban politics. In writing (and drawing and painting and postering and stickering and knitting and scratching ...) on the surfaces of the city without the permission of authorities, graffiti writers and artists act as though they have a right to write which they do not have according to the police order of the city. In acting this way, graffiti writers and artists stage the 'double movement of exclusion and inclusion' which is at the heart of politics in Rancière's formulation (2006: 61). This double movement is illustrated by the quotation from Banksy used to open this section.[2] He asks us to imagine a city 'where everybody could draw wherever they liked', a city that 'belonged to everybody'. Here, a claim is being made on behalf of 'everybody' against authority, where 'everybody' stands in for anyone who occupies or moves through urban space (although we will reconsider the relationship of graffiti to authority in the next section). In Rancière's terms, a claim is being made on behalf of the part of the city which has no part in the urban media landscape according to the existing police order. And yet, as the last line makes clear, neither Banksy nor other graffiti writers and artists are prepared to wait for this claim to be recognized by authorities. The walls are already wet with paint, precisely because at least some people are already acting to make this imagined city a reality by getting on with it.

Of course, as Rancière observes, politics is always confronted with and *productive of* police. Graffiti's challenge to the authority of the authorities is met with resistance, through various efforts designed to limit writing and drawing to its proper authors and its proper places in the city. But the policing of graffiti is, in fact, more complicated than it may at first appear. Notwithstanding what I have just argued, we should be wary of simplistically characterizing the policing of graffiti as a contest between 'the graffiti writers and artists' and 'the authorities'. In the next two sections, I want to elaborate on this claim, in order to illustrate some wider arguments about the nature of policing in cities. First, I will demonstrate that the policing of graffiti is not solely a matter for the (uniformed) police, but in fact involves a range of actors who mobilize competing claims to authority and techniques of control. Second, I will argue that graffiti writers and artists are not simply the targets of policing efforts by others – rather, they are themselves engaged in efforts to police graffiti.

'A bit of vandalism here, a bit of graffiti there ...'

> In isolation a bit of vandalism here or a bit of graffiti there might seem trivial, but their combined effect can seriously undermine local quality of life. Some criminologists talk of the 'broken window' problem. They argue that a failure to tackle small-scale problems can lead to serious

crime and environmental blight. Streets that are dirty and threatening deter people from going out. They signal that the community has lost interest. As a result, anti-social behaviour and more serious criminality may take root. (UK Prime Minister Tony Blair, 2001)

In recent years, the growth and mutation of graffiti has emerged as a matter of considerable concern for policy-makers and the police. In many cities, graffiti has been characterized variously as a 'quality-of-life offence' and/or a form of 'anti-social behaviour' which unsettles 'the community' by undermining the moral order of cities. Graffiti and those who write it have become targets of the police in uniform. But a much wider range of actors have made the placement of graffiti their concern – from private security guards and property owners to urban planners and architects, youth workers and advocates, art collectors and critics, not to mention corporations and senior graffiti writers themselves.

In order to understand the similarities between these apparently disparate policing efforts, it is necessary to consider the policing of graffiti more broadly. Drawing on the politics/police framework developed above, I want to argue that different ways of policing of graffiti are founded upon a police order which denies graffiti writers and graffiti any part in the city (as both physical space and polity).

In developing his understanding of police, Rancière argues that:

> The police is ... first an order of bodies that defines the allocation of ways of doing, ways of being, and ways of saying, and sees that those bodies are assigned by name to a particular place and task; it is an order of the visible and the sayable that sees that a particular activity is visible and another is not, that this speech is understood as discourse and another as noise. (1999: 29)

Now, graffiti is of course highly visible – that is its point! But while it might be visible, in the police order of the city graffiti is not registered as a claim on the city, but rather as a kind of 'background noise' which has nothing to *say* to or about the city. In Rancière's terms, the police order of the city deprives graffiti writers of logos, of meaning, of what he calls 'symbolic enrolment in the city' (1999: 23). We can see this police logic at work in a variety of different policing operations.

Let us consider first the most common policy/police response to graffiti, which I will call here the *repressive* response. In a growing number of cities, politicians and policy-makers have declared war on graffiti. These wars on graffiti are closely associated with the emergence of operational priorities and procedures adopted by police forces. Paradigmatically, the original growth of new forms of graffiti in New York City in the 1970s and 1980s was one of the key factors that gave rise to new approaches to law (with the criminalization of so-called 'quality of life' offences such as graffiti and pan-handling) and order (with the application of so-called 'zero tolerance' policing towards these new offences) (Austin, 2001; Dickinson, 2008). Variations on this thesis have gone on to become incredibly influential within and far beyond New York City. As evidenced by the quotation from UK Prime Minister Tony Blair above, by the early 2000s the broken windows or 'quality of life' agenda had spread to the UK, where it had been incorporated into the government's new Anti-Social Behaviour policies.

In places where 'quality of life' has become a priority for policing, penalties for graffiti writing, and frequently also for possession and sale of graffiti writing implements, have increased substantially. Police have also been given new powers to search, fine and arrest graffiti writers, and specialist graffiti squads have been established within police forces to gather intelligence as part of the wider zero tolerance strategy which informs the 'wars on graffiti' (Iveson, 2010).

But increasingly, the war on graffiti is not only a matter for uniformed police. Private property owners and communities have been encouraged, incentivized and sometimes compelled to 'take responsibility' for graffiti prevention and removal, as part of a wider agenda to distribute responsibilities for the maintenance of public order (Garland, 2001). This trend helps to explain the recent enthusiasm for 'Crime Prevention Through Environmental Design', which has given urban designers and architects a starring role in efforts to police the city. Design has been advocated as a cheaper and more effective method for controlling the ways in which written media can be applied to urban surfaces (among other things). Surveillance, lighting, sightlines, so-called 'green screens' such as hedges and shrubs, textured and graffiti-proof building materials and even public art have all been suggested as measures that might help to 'design out' graffiti.

Where graffiti persists, a growing graffiti removal industry is ready to step in to help property owners and urban authorities maintain the propriety of urban surfaces through 'rapid removal'. The City of Sydney Council, for example, has engaged the services of a company named Techni-Clean to implement its graffiti management policies – 'Keeping Cities Cleaner' is the motto emblazoned on the trucks and work-shirts of the Techni-Clean street teams, a perfect illustration of the police reading of graffiti as nothing more dirt (Cresswell, 1996).

This repressive approach to the policing of graffiti is clearly premised on a particular police order of the city in which graffiti has no place. Those who write graffiti are characterized as 'anti-social', as 'mindless vandals', as 'gangsters', as 'egotists', as 'idiots'. No account is made of their writing and art. At worst, their graffiti is a 'nuisance', it is 'meaningless scribble', it is 'formless splashes of colour', it makes places 'dirty'. At best, even if it appears to 'say something' or is 'beautiful' or 'artistic', graffiti is by definition seen to be *misplaced*. As such, graffiti writers have no part in the community, because they have refused to take their part by obeying the law and respecting other people's property.

Of course, not everyone views graffiti in this way. For instance, in many cities repressive graffiti policies are contested by those who argue for an approach that is sometimes referred to as 'harm minimization'. Advocates of this approach seek to recognize and then redirect the energies of graffiti writers by providing legal opportunities to develop and practise their art (Iveson, 2007). Here, graffiti is not so much to be repressed as *re-placed*, through the expanded provision of recognized public art programmes and spaces which give it a proper place in the city, and through broader policies of social inclusion and cohesion which improve the lives of young people.

But to reject this repressive approach is not *necessarily* to reject the police order of the city in which graffiti remains a product of outsiders who are deprived of meaning. There are variations on this theme. For instance, some advocates of the harm minimization approach towards graffiti reject the repressive approach on the grounds that graffiti is written by the 'deprived', the 'marginalized', the 'misunderstood', the 'alienated' who have been denied a proper place in the community. But for those who view graffiti writers in this way, graffiti sometimes still only registers as 'the noise of aggravated bodies' (Rancière, 1999: 53), a kind of anonymous cry of pain rather than a *political* discourse. Others who embrace the harm minimization approach do so in order to provide opportunities for graffiti writers to 'make the right choice' and 'be responsible' by writing and painting legally and not illegally – here, the aesthetic, cultural and even economic qualities of graffiti-style writing and art are recognized, but once again this recognition is premised upon graffiti writers choosing to accept their authorized, proper place in an orderly urban media landscape. For those that don't make the right choice, repression remains justified.

Now, in characterizing both repression and harm minimization approaches to graffiti as examples of policing, I do not wish to minimize their very important differences. But given that most policy debates over graffiti tend to be dominated by these two positions, which are presented as opposing views, I think Rancière's understanding of 'the police' is useful precisely in identifying their *similarities* as well as their differences. Both seek a way to manage the 'graffiti problem' in order to uphold the natural order of the city, in which property ownership and planning law dictate the possibilities of the urban media landscape. Graffiti is a problem that needs to be managed for the sake of good order – this is a far cry from the vision of the 'graffiti problem' articulated by Banksy above. So, while we might indeed wish to argue that one form of policing is better than another (a point to which I will return in the conclusion), as Rancière asserts in the quotation with which I began this chapter, that does not make them any less the opposite of politics.

But we should not stop our discussion of the policing of graffiti here. For alongside politicians, the uniformed police, architects and urban designers, criminologists and youth workers, street-cleaning contractors and technology developers, there is another group of folks who actively engage in efforts to police the activity of graffiti writers …

'You suck until further notice ...'

> There are a few things you must do before in order to make your presence a welcome one. First; know the history. Second; know the rules of the game. Third; work hard at being good, or are least competent. Fourth; snitches, and shit talkers get stitches and need walkers. Fifth; you're good, but you're not that good. Keep your fat head to a reasonable swell and get back to work. (Mark Surface, 'So You Wanna Write on Walls')

One of the strengths of Rancière's approach to the police is that it refuses to align either the police or politics with a particular subject. As such, we should not consider the policing of graffiti in the city as simply a matter of 'the authorities' versus 'the graffiti writers'. Rather, efforts to control the placement and form of graffiti are also enacted by graffiti writers and artists themselves.

Stephen 'Espo' Powers, in his (1999) book *The Art of Getting Over: Graffiti at the End of the Millennium*, tells a story about coming across a document called 'So You Wanna Write on Walls', stuck on the wall of the Graffiti Writers Local 132 Union Hall. As he tells is: 'At first, its size and format made me think it was a religious tract. Upon closer inspection I realized its purpose was to instruct on another level. I submit it to you without any edits, so read carefully, you saps.'

First among the rules listed in this set of instructions credited to 'Mark Surface' is: 'you suck until further notice'. For the aspiring graffiti writer, the rest of the rules provide some tips about how to suck less in the eyes of your peers. The document provides some tips on how to choose a good tag name, and how to develop an individual style. There are also some quite specific rules about where to write, which include: 'Fuck permission walls, write your name bigger every time you go out ... Don't write on houses of worship, people's houses in general, other writers names and tombstones. Writing on memorial walls and cars is death.'[3]

Now, I do not want to elevate the status of this document, to present it as something that it is not. It's meant to be funny – there is no graffiti writers' guild or union, which limits entry to the profession and enforces a code of conduct! No doubt most people who write graffiti do so in complete ignorance of this particular document. But I think 'So You Wanna Write on Walls' also speaks to something significant for our discussion of policing. There certainly *are* graffiti writing counter-publics, through which aesthetic criteria, ethical codes of conduct, and even authority and hierarchy are discussed and debated (Iveson, 2007). Indeed the widespread reproduction of this document in magazines and on the internet is evidence of such counterpublics in action, where 'style', the 'rules of the game', and respect for the 'old school' and 'kings' are frequently discussed.

The aesthetic criteria, codes of conduct and claims to authority that are debated in these counter-publics are sometimes connected with efforts to *police* the writing of graffiti, when they are translated into rules about what is 'proper' and 'improper' graffiti. So, for instance, within graffiti writing scenes there are tensions between those who value 'quality' and those who value 'quantity' – should writers be focused on producing the most stylistic pieces, or on carpet bombing the city with serial reproductions of tags and throw-ups in as many places as possible? Are train yards proper places for girls? Growing rifts between some personalities associated with the 'graffiti' and 'street art' scenes have also emerged based on disagreements about the codes of the street. A high-profile example of this disagreement between codes of the street, or to put in it Rancière's terms, forms of *policing*, is the recent 'beef' between graffiti artists Banksy and Robbo in the UK. This feud has revolved around the stencil artist Banksy being accused of disrespecting the painted work of Robbo, an acknowledged 'king' of more traditional hip hop graffiti in London (see Steinhauser, 2010). Figure 5.2 provides an example of how this 'beef' has played out on the walls of London. In this image, Banksy has stencil painted over an existing Robbo piece, by literally depicting a decorator in the process of covering up the previous Robbo graffiti. Whilst the specific representational meaning intended by Banksy might be up for debate (e.g. Banksy is signifying a replacement of 'older' forms of graffiti with new, more ironic forms, or Banksy might simply be seeking to erase Robbo's artistic presence from the London landscape). Whatever the intended or interpreted content, the feud and its manifestations in street art (see

Figure 5.2 The infamous Banksy piece which partially covered a 25-year-old piece by Robbo in London. Photo by Matt Brown. (source: www.flickr.com/photos/londonmatt/4202570401/)

Figure 5.2) demonstrate how a diverse graffiti community, which might be thought of as outside of policed orders, actually has its own orders and related conflicts.

Of course, the priorities and procedures of policing employed within the graffiti writing scene are very different from those employed by the urban authorities discussed above. Contravention of 'the rules' might result in your pieces being 'capped' by another writer, in threats of violence, or in your name being dissed in a magazine or an online forum. But notwithstanding these differences, we can still call these aesthetic criteria and codes of conduct a form of policing. So, while above I asserted that graffiti constitutes a form of politics in relation to the police ordering of the urban media landscape, here I want to qualify that assertion, by insisting that graffiti is not pure politics. Graffiti writers might well insist that they are the equals of the authorities, that they have the right to write on the city – but many of them also frequently insist that they are not equals with each other!

Conclusion

> From Athens in the fifth century B.C. up until our own governments, the party of the rich has only ever said one thing, which is most precisely the negation of politics: there is no part of those who have no part. (Rancière, *Disagreement*, 1999: 14)

In this chapter, I have sought to make the case for a particular way of understanding the relationship between urban politics and police, illustrating this approach with an extended example of the policing of graffiti.

The policing of graffiti, as we have seen, takes a number of forms. These include attempts to repress graffiti, or to minimize it and contain it in its proper places. A variety of actors are involved in these policing efforts, including but not limited to the uniformed police. But these policing efforts share a crucial characteristic that makes them all 'police' in the sense I have used in this chapter. They all attempt to *naturalize* a particular form of authority over urban surfaces which denies and/or restricts graffiti writers a place on those surfaces. Those who pursue the repressive and harm minimization responses to graffiti attempt to naturalize property relations in the city, insisting that surfaces are inviolable because they are the property of private or statutory authorities. In their city, access to these surfaces is either to be bought or sold, or planned for – it is not to be taken on the basis of inhabitance and physical access. For graffiti writers engaged in their own efforts to police the surfaces of the city, there is an attempt to naturalize certain surfaces as the 'property' of established graffiti artists, and access is to be earned by following the rules and the acquisition of subcultural capital.

These forms of policing of graffiti not only work by restricting access to surfaces, then, but also through the *account that is given* of graffiti. The naturalization of authority works in part through the reduction of graffiti writers to people who write but have nothing to say. Their writing is not recognized as a form of speech, but is characterized as 'meaningless scribble' or 'vandalism' or 'splashes of colour' or 'crap' which has no place/part in the city. As Rancière argues, this is the defining characteristic of the police: 'If there is someone you do not wish to recognize as a political being, you begin by not seeing them as the bearers of politicalness, by not understanding what they say, by not hearing that it is an utterance coming out of their mouths' (2001: 23).

To put this another way, for those policing the surfaces of the city, unauthorized graffiti writers do not have anything to say, even though their graffiti marks its many surfaces. There is no point speaking to them about what they do, because they do not speak properly – their graffiti is unauthorized, and as such its authors do not have a place in the city.

The understanding of the policing of graffiti I have developed in this chapter implies a related understanding of the graffiti as politics. Attempts to politicize the urban media landscape through graffiti are not just a matter of writing graffiti. A politics of graffiti must also involve the assertion of a different account of graffiti that insists that graffiti writers too are part of the city, even as they are denied authority. The quotation from Banksy that I used earlier in this chapter is one example of a graffiti politics. Another example comes from Sydney graffiti writer Pudl, whom I interviewed as part of a campaign to shift graffiti policy from repression to harm minimization in Sydney in which I was involved. In reply to a question about his ideal graffiti policy, Pudl replied: 'There should be no graffiti policy. The government cannot legislate against an artform.' As he went on to explain, to him the whole point of graffiti is that it challenges the 'natural order' of urban public space: 'graffiti is a process, as much as it is a finished product. The philosophy behind this process stems from making private spaces public and challenging the rules of access whether it be physical access or social access. Pure vandalism at its core is a political act as much as it is an artistic image.'

Other artists have similarly insisted on the political dimension of graffiti, albeit in different ways. Espo has now resolved to write graffiti in broad daylight – while he still often conducts his work without permission, his refusal to hide is his way of declaring that he is doing nothing wrong (Powers and Vanetti, 2010). Jordan Seiler, a New York artist who replaces outdoor advertising with his own art works, takes a similar approach, and he has instigated a number of wider campaigns to reclaim outdoor media for public communication.[4] All of these artists insist that there is nothing wrong with what they are doing, and they talk, blog, write and speak about their graffiti and art as part of the city, something that makes the city better.

In reflecting more broadly on policing, I want to draw one final point out of this extended example. Hopefully, I have demonstrated the usefulness of a particular approach to 'the police' for a wider consideration of urban politics. The key point here is that if we wish to understand how the city is policed, it is not enough to ask what the petty/uniformed police are up to. This is not just a point about how the contemporary 'cultures of control' extend well beyond the police themselves to a range of other actors. It is a point about the constitution of the city itself as a political space. To ask about how the city is policed is to ask about how some ways of ordering spaces and their possibilities are *naturalized*. And so, to ask about urban politics is to ask about how those who have no part in this orderly city insist on its contingency and assert that in fact they do have a part, based on a declaration of their rights to inhabit the city as equals.

Notes

1 For the moment, I will leave aside the question of where 'policy' fits into this schema. Here, I will observe only that the *Oxford English Dictionary* tells us that 'In early use [policy was] sometimes indistinguishable from police' and that 'a number of the senses of French *police* are represented more commonly in English by policy'. As we shall see, Rancière's use of the word 'police' would appear to be consistent with this observation.
2 British street artist Banksy is currently one of the world's most (in)famous street artists. See www.banksy.co.uk for a sample of his work.
3 You can access the whole document at www.hiphop-network.com/articles/graffitiarticles/soyouwannawriteonwalls.asp.
4 See www.publicadcampaign.com.

References

Austin, Joe (2001) *Taking the Train: How graffiti became an urban crisis in New York City*. Columbia University Press: New York.
Banksy (2000) *Existencilism*. Weapons of Mass Destruction: London.
Blair, Tony (2001) *Improving Your Local Environment*. Speech to Groundwork Trust, Croydon. Available at www.number-10.gov.uk/output/Page1588.asp (accessed 9 January 2006).
Cresswell, Tim (1996) *In Place/Out of Place: Geography, ideology, and transgression*. University of Minnesota Press: Minneapolis, MN.
Dickinson, M. (2008) 'The making of space, race and place: New York City's war on graffiti, 1970–the present', *Critique of Anthropology*, 28(1): 27–45.
Dikeç, M. (2005) 'Space, politics, and the political', *Environment and Planning D: Society and Space*, 23(2): 171–188.

Garland, David (2001) *The Culture of Control: Crime and social order in contemporary society*. Oxford University Press: Oxford.

Hermer, J. and Hunt, A. (1996) 'Official graffiti of the everyday', *Law and Society Review*, 30(3): 455–480.

Iveson, K. (2007) *Publics and the City, RGS-IBG Book Series*. Wiley-Blackwell: Oxford.

Iveson, K. (2010) 'The wars on graffiti and the new military urbanism', *City* 14(1–2): 115–134.

Powers, Stephen (1999) *The Art of Getting Over: Graffiti at the millenium*. St Martin's Press: New York.

Powers, Steve 'Espo' and Vanetti, Serena (2010) 'Steve ESPO Powers: writers Writing about writing', *Very Nearly Almost* 13: 20–27.

Rancière, Jacques (1999) *Disagreement: Politics and philosophy*. Translated by J. Rose. University of Minnesota Press: Minneapolis, MN.

Rancière, Jacques (2001) 'Ten theses on politics', *Theory and Event* 5(3): 12–16.

Rancière, Jacques (2006) *Hatred of Democracy*. Verso: London.

Rancière, Jacques (2009) 'Afterword: the method of equality, an answer to some questions', in Rockhill, G. and Watts, P. (eds) *Jacques Rancière: History, politics, aesthetics*. Duke University Press: Durham, NC. pp. 273–288.

Steinhauser, G. (2010) 'A game of tag breaks out between London's graffiti elite', *Wall Street Journal*, 3 March.

6

MAYORS AND THE REPRESENTATION OF URBAN POLITICS

Donald McNeill

Introduction

Mayors occupy an enigmatic position within accounts of urban politics. By its nature the institution of mayor underlines the significance of an individualized agent, chosen directly as an executive leader by the electorate, or else indirectly voted into office by elected councillors, usually representing the dominant political party. Whichever route is chosen, it should be apparent that mayors will tend to be individuals with the political skills to unite diverse party factions, or else appeal to a broad cross-section of the urban electorate.

While many accounts of mayoral governance, particularly in the political science literature, tend to focus on the mayoralty as an institution, my preference here is to consider the mayor as an individual, as a personage, who performs a mayoral role. From this perspective, the individual enrols a range of actors with diverse claims on political power (from businesses and interest groups which fund their campaigns, to 'fixers' who deliver their political deals). Above all, they perform the function of representation – but in a double sense. On the one hand, they are elected (chosen, etymologically) by voters to ostensibly re-present their views, in a delegated form of power. On the other, they perform acts of 'representation' of a particular set of worldviews and scenarios for the elected spatial formation that they represent. As Norman Fairclough has identified in his study of British Prime Minister Tony Blair, such politicians are 'on the one hand speaking with impersonal authority', or on behalf of the 'world-community', about what is the case (epistemic modality), what will be (predictions), what should be (deontic modalities), yet on the other hand speaking personally ('I' statements) and on behalf of an inclusive 'we' – community of common experience ('we all')' (Fairclough, 2003: 181). Via speeches, policy statements, their minuted interventions on particular policy debates before council, their electoral campaigning leaflets and media broadcasts, mayors also represent 'the state of the city'.

Current debates on the nature of urban government have emphasized the structural power of the state, especially its territorialization through legal, regulatory and economic powers, with less emphasis given to political agency in many accounts (often dismissed as 'voluntarism' within certain, particularly neo-Marxist, theoretical traditions). A central theme of this work has been the plausible assertion that state power has been reconfigured in ways in which a 'vertical' understanding of political scale, based on hierarchy, is no longer tenable. Instead, others have argued for a network perspective, based on lateral, distanciated engagements with political and economic actors elsewhere. Others still have argued for a topological approach, which might be characterized as the study of the 'reach' of state power (Allen, 2010; Allen and Cochrane, 2010).

While mayoral powers vary from city to city, there is nonetheless a case for locating a strong degree of urban political power in the hands of this one individual. The chapter provides a brief introduction to how mayors play a representational role in political life in the dual sense noted above. It begins by addressing the spatiality of mayoral leadership, highlighting ways in which mayors claim the territories which they have been elected to represent. This is done through reference to the aestheticization of politics, where place symbols and historical narrative form important elements in the political discourse. Second, the chapter flags the importance of seeing mayors both as embodied institutional forms, as well as nodes in party political networks. Following Low (2007), it suggests that an understanding of party politics is essential to any theorization of mayoral power in contemporary society. Third, the chapter briefly considers ways in which mayors are involved in the governmentality of cities, setting out agendas that shape 'moral geographies'.

Mayors and Political Territory

An important distinction can be drawn between political geographies of mayoral leadership and the accounts of mayoral power that are orthodox within political science and urban sociology. To begin, I want to sketch out four crude notions of mayoral geographies. First, there is the question of scale. There has been significant debate among geographers over how best to conceptualize *scale*, as it is often conventionally understood as being related to size, with a 'nested hierarchy' of central government, regional or state government, then city or municipal government. In this framework, cities have the least power and autonomy. In some ways, this crude distinction has merit – there is certainly little to compare between the treasuries of most central governments and those of their cities. However, once geographical differentiation is brought in, it is clear that comparing the power of the mayor of the nation's capital city with that of his or her provincial counterparts is likely to give their respective politicians a very differentiated access to the networks – financial, cultural, media, political – that coalesce around capital cities.

Second, the importance of orthodox Cartesian notions of territory become very important in the analysis of mayoral politics. Here, the *electoral geographies* of mayoral power can refer to the need for mayors to win elections district by district, street by street. In reality, this is true for all forms of political activity, except that in the magnified forum of the city, mayors have historically been reliant on securing support down to the level of wards, or even individual polling booths. Notoriously, this can extend to the appointment of 'ward captains', small-scale power-brokers capable of delivering block votes in return for public service jobs, new local facilities, or some other form of political gift. The early examples of this in the United States – Tammany Hall in late nineteenth-century New York, or the Mayor Daley machine that controlled Chicago during the 1960s and 1970s – remain some of the most noted examples of the micro-spatial underpinning of mayoral power.

Third, leaving aside the often vexed issue of how far mayors favour particular neighbourhoods within cities, it is important to emphasize that mayoral electoral success often revolves around who best tells a convincing story about the city as a unified community. This could be referred to as a *representational space* (cf. Lefèbvre, 1991), a performing and practising of political activities that shape material space. As I describe elsewhere, Barcelona's Maragall (1982–1997) was particularly adept at forging a strong support based on his ability to engage an 'imagined community' of Barcelonans (McNeill, 2001b). So, for a time, was Rudy Giuliani able to do for New Yorkers (McNamara and McNeill, 2012) The basis for such 'epistemic modalities' (Fairclough, 2003) depended upon a centrist reading of the city's history, one that laid the basis for a future-oriented set of policy goals. In turn, this required the creation of a sense of the place of the city within a wider set of political relationships – in the case of Maragall, with Spain and with Europe. By contrast, in the case of Giuliani, there was a focus on the 'enemy within', with those who were deemed not to share in the particular class-based political vision that Giuliani had in mind for the city. This played into his 'moral geographies' of New York, as I discuss below.

Fourth, we can refer to the *distanciated geographies* of mayoral politics. For example, it has been suggested that just as nation-states have foreign policies, so should city councils. As such, a relational politics of transnational collaboration offers a fruitful political forum for international solidarity. An example came in the controversial – though largely gestural – exchange orchestrated by Ken Livingstone while Mayor of London, with Hugo Chávez, the Venezuelan President, in the form of subsidized oil for public transport, in return for policy advice (Massey, 2011). However, the importance of these forays is often overstated.

These four elements of mayoral geographies – the size of the mayoral treasury relative to other actors, the impulse for electoral control, the desire to represent a political space, and the extra – local networks and relations that cities engage in – form the backdrop to contemporary debates on mayoral power.

Bringing together these elements became increasingly, necessary for urban scholars as the field of urban governance became more complex. This was especially so from the 1990s, as local authorities began to look at 'creative' ways of financing both their capital costs (particularly spending on electorally popular projects such as transport systems, arts and culture venues, and sports stadia) and, even, on their recurrent budgets devoted to paying for their staff and services:

> During the Millennial boom of the late 1990s and the first decade of the 2000s, many U.S. municipal governments acted as the kind of innovators that we associate with private market actors. They assisted capital in finding profitable investment outlets by devising new ways to monetize their own assets and create new securities. They turned income streams from their existing and future tax bases, infrastructure, and pension funds into fungible securities and helped build secondary markets for their exchange, both of which momentarily took some pressure off of the global capital glut. (Weber, 2010: 270)

In the immediate aftermath of the global financial crisis, it became clear that, around the world, a large number of city councils and municipalities had lost significant sums of money through betting on derivatives and other financial instruments that had lost value in the crash.

This has been an intriguing moment in the study of urban politics. One of the most influential texts in urban politics theory in recent times has been David Harvey's (1989) development of the term urban entrepreneurialism. Taking as his starting point the book *Urban Fortunes: The Political Economy of Place*, by Logan and Molotch (1987, 2007) Harvey provided a spatial framework for interpreting how urban growth coalitions worked with spatial division of consumption, production, and government financing. What unites both accounts is the way in which city councils are seen to be acting as if they were private firms, operating in a marketplace where their city is the product. What has become very interesting is the way in which mayors, along with urban policy-makers more generally, have taken advantage of improved and intensified processes of knowledge circulation to inspect, borrow and copy elements of urban policy solutions elsewhere (McCann, 2011; McFarlane, 2011).

There are numerous organizations which bring together mayors from different cities to lobby central government, or share ideas, such as the think-tank based in London, City Mayors. While this kind of policy learning can sometimes appear to be voluntary, it should also be noted that an important set of globally operative financial actors are active in lobbying city hall to commission particular products. Katz (2010) documents the extreme case of the mayor of Saint Étienne, Michel Thiollière, who as early as 2001 was buying the now notoriously toxic swaps and derivatives, such as the collateralized debt obligations which local authorities around the world bought from door-to-door salesman/investment bankers such as Goldman Sachs. By 2008, when the world financial system entered its credit crunch, Saint Étienne found itself with huge debts based on the speculative nature of the contracts they had signed, putting them in huge debt with the banks. This story has many intricacies: what is important here is that in the elected mayor is signing the contract on behalf of the municipality she or he is elected to represent. When they leave office, they leave the contract and the financial commitments behind. Interestingly, as Katz (2010) notes, Thiollière had in 2004 been named 'best mayor in the world' by the City Mayors think-tank. By the end of the 2000s, he was not seen in such glowing terms.

> **BOX 6.1 MAYORS AND SOCIAL MEDIA**
>
> The increasing importance of social media within political life can have significant consequences for how mayors communicate with the public. Michael Bloomberg, New York's long-serving mayor, has been one of the most proactive adopters of social media as a means of communicating his policy ideas, promoting New York as a world-leading digital city, and maintaining a consistently high public profile through the likes of Twitter and Foursquare, alongside more conventional media communications strategies. However, Grynbaum (2012) describes how Bloomberg has began to acknowledge that this can have its downside, reporting on a speech in Singapore where he suggested that 'Social media is going to make it even more difficult to make long-term investments in cities'. By this Bloomberg meant that every policy issue that he had announced generated a substantial degree of online commentary and criticism. During the speech Bloomberg continued: 'We are basically having a referendum on every single thing that we do every day.' By this he is presumably suggesting that the time-frames used by the public (usually the four-yearly election event) to judge mayoral abilities have shrunk markedly to an almost daily diagnosis of policy. While Bloomberg is undoubtedly exaggerating for effect, he identifies the challenges posed by social media to the conventional political process. The opportunities offered to 'council watchers' to set up oppositional websites or blogs to crowdsource information on failing council services have certainly taken the urban informational 'regime' to a level hitherto unforeseen. And for mayors, who require a degree of personal popularity to gain widespread public support, they require a careful assessment of how far to invest in social media – in both time and money – as opposed to keeping a low-key presence in the social media sphere.

Party Politics, and the Mayor as Spatial Actor

An argument advanced in support of the strong model of mayoral governance, where voters can directly choose an individual rather than a party, is that it can allow the emergence of politicians capable of evading intra-party factionalism and inter-party oppositionalism. Both forms of government – especially when tied into the temporality of electoral cycles – are seen as leading to irrational, inefficient decisions regarding the allocation of public funds. This view was particularly active around the mid-1990s when under UK Prime Minister Tony Blair legislation was introduced to allow cities to introduce directly elected mayors. For Blair, whose political approach involved the increasing involvement of business in public life, an ideal candidate for mayor of London would have been the multi-millionaire Richard Branson. Low (2007) summarizes this position well:

> One argument for supplanting party-structured councils with relatively unencumbered executive mayors, common in media debates about urban politics, is that, literally, mayors are better or more 'proper' actors than are collectivities of politicians. Mayors can get things done, decide decisively, and provide a basis for popular identification and hence better accountability. Read quickly, these sorts of arguments sometimes seem rather convincing at an empirical level, but they embody various normative emphases about democracy that are worth discussion, particularly as post-party-politics arguments have been important in talk about the virtues of executive mayors in the UK and elsewhere. (2662)

As Low notes, such directly elected mayors were a popular, if naïve (or indeed disingenuous), response to the growing demands for mayoral politics in UK cities in the early 2000s. In reality, mayoral politics are less often about 'apolitical', charismatic individuals, than about time-serving politicians who have gained great influence within their chosen political party. This is rarely mentioned in much of the scholarly mayoral literature. As Low continues:

> The scarcity of theoretical literature on political parties as opposed to social movements, business coalitions, and other actors in cities is puzzling. Although in different contexts, parties have a variety of forms, degrees of organisational or ideological coherence, and are differentiated in terms of structure, strategy, goals, and membership, they do permeate, however incompletely, many prereflective conceptions of what contemporary 'politics' is about. (2007: 2652)

With this in mind, we can surmise that autonomous mayoral power may be overstated. Even New York's Bloomberg mayoralties (2001–2013), widely seen as an exemplification of non-partisan, independently funded individual politician, were in reality given impetus by a network of Republican Party advisors, sympathizers, and activists who both secured his electoral victories, and enacted his policies in office.

Indeed, many theories of representative politics have suggested that its nature is defined by the seizure of the state's tools by power blocks operating via political party systems. As such, party (rooted in civil society) and state become very close. For example, in China, as with most state communist systems, there is a very narrow separation between the political party (the Chinese Communist Party) and the state institution (the municipality of Beijing, say). In reality, the economic and political significance of Shanghai (as China's financial leader) and Beijing (as its government centre) began to structure Chinese politics at the central government level. In each case, the power of the local party chief was greater than that of the city mayor, although both were likely to be well positioned within the city's political hierarchy. The Chinese experience has certain parallels with that of the US, as Logan and Molotch suggest, in how both reflect the importance of political careerism:

> Chinese local politicians don't need to curry campaign contributions from real estate entrepreneurs ... But the national government evaluates local officials by the criterion of how large they have grown their economies. Career paths follow accordingly; officials can move up to the next governmental tier, with commensurate gains in wealth and standing, based on their performance. The trick is to generate infrastructure and investment from autonomous, state-owned enterprises and the private sector to grow the locality. (2007: xvi–xvii)

This process – of using the networks engendered among other politicians and major economic institutions – is not confined to China. In many countries, becoming mayor of a major city can also be a springboard to higher office, given the electoral visibility it can allow, as well as the embedding within regional or national party hierarchies. This can be conceptualized as 'jumping scales', using the territorial power base of a large municipality as a demonstration of political aptitude (not just to the electorate, but also to party colleagues). Alternatively, using a different metaphor, it can involve skills in 'networking' which includes the ability to develop distanciated relationships with other political allies who may have a power base many hundreds of miles away in the same nation-state. This is where the political party is an essential adjunct to the territorially bounded nature of municipal 'government'. As Low argues:

> In joining up very different citizen constituencies across space (and via symbolic tradition, through time), maintaining relations with business and other sources of resources, connecting with a variety of other political entities, including movements of various kinds, and linking all these with state institutions and policy debates, formulation, and implementation, we could say that parties have been exemplary forms of governance structures before governance as such was ever said to be 'rising'. (2007: 2661)

Thus taking political parties seriously is an important element in understanding the spatial constitution of urban power, particularly as it is manifested through ambitious political actors. In the case of mayors, this may even indicate a degree of territorial entrapment, where their overly powerful association with a particular city may not play out well when running for higher territorial office. For example, Rudy Giuliani's often strained relationship with the New York Republican Party and limited voter base were evident in his failed bid for the Republican Presidential nomination in 2008, perhaps illustrative of too close a personification of the city, where a chosen individual both represents, embodies and narrates a particular understanding of the city, and contributes to its moral geographies.

The Mayor and Moral Geographies of Urban Life

In this section, I consider the significance of governmentality in the context of city leadership. This body of work, originating from the French social theorist Michel Foucault, but developed by a number of others (see, for example, Joyce, 2003), has focused on the rise of government as a set of skills, or technologies, which have sought to expand the state's capacity to deal with the rising populations of contemporary cities. This school of thought is quite different from Marxian approaches to urban governance, focusing more on how the modern political subject is monitored, observed, encouraged to self-organize through various forms of acceptable conduct (e.g. to pay taxes, to not drop litter, to send children to school, to work hard, and so on).

This is important because what we think of as 'cities' are not always mapped neatly onto political units. A common complaint of mayors, even directed to colleagues in their own political parties, is that the large contemporary city is a unique spatial formation, posing very different challenges to running a nation state. This is for a number of reasons. Most importantly, city councils are notoriously underfunded, because the claims to their territory are not synonymous with those who pay taxes or rates for its services. A common feature of North American government, for example, is the issue of suburban secession, where small, affluent communities break away from the territory where their employment is located. This may mean that the costs of providing car parking, road maintenance, street cleaning, and so on to house so many commuters is not matched by tax revenues, notwithstanding corporate taxes.

The role of mayors in the context of governmentality should be evident. Mayors have a very immediate influence on how the city's residents behave, via the enactment of various policies designed to influence the performance of everyday life. This is often subsumed under a rubric of morality, and is often related to how the city's public spaces are regulated and organized. It is very clear, looking from city to city, that some cities crack down on street busking more than others, for example, or are more or less tolerant of homeless people sleeping in the streets, or may permit people to drink alcohol only at café tables, or allow particular forms of visual or bodily display. These activities are often related to a set of moral precepts; they may be associated with particular modes of religious identification. For example, some may regard sitting in a park drinking a bottle of beer to be a relaxed form of urban space use, associated with responsible citizens capable of regulating their own intake of alcohol, and with an assumption that such activity is unrelated to children drinking at an early age. Other schools of thought may see this as being a very different issue, regarding tougher standards of public decorum.

During Rudy Giuliani's period as mayor of New York (1993–2001) we saw an extreme example of this. There are many who associate Giuliani with his activities on and after 9/11, when he was widely admired with providing effective leadership on the city's streets at a time of great trauma and uncertainty. However, it is argued that 'in the glow of all the attention, his mayoralty came to be cast, inaccurately, as some kind of model of civic governance' (Polner, 2005: xxxv). What Giuliani did with great skill was to link broader discourses of morality with specific iconic spaces within the city. He singled out Times Square, for example, as an example of how lawlessness (often associated with particular ethnic groups within the city) had to be eradicated, rather than the equally organized white-collar criminality that was unravelling around him. This allowed him to build political legitimacy for an increasingly controversial backing of police brutality (McNamara and McNeill, 2012).

This relationship to territory is *performed*, in three inter-related roles. First, mayors 'embody' cities. Many are born in the city they represent (and here representation has a double meaning), and will have some sort of relationship to the essentialized characteristics of its inhabitants (often accent, sense of humour, or 'inheritance' of attributes of an idealized predecessor). Here, they communicate with their voters/fellow citizens. Second, they will act as animator of city space. Rather than pursuing an abstract notion of territory, mayors often strive for visibility in the everyday life of the city, especially at times of crisis. Finally, they are likely to provide some sort of narrative in their press conferences and public appearances about the immediate past, present and future of the territory they represent. By extension, they will seek narrative coherence in this role (the 'what is the case, what will be, what should be' that Fairclough identifies above), and are likely to perform a public discourse concerning crime, fear of terrorism, the economic climate, and how this relates to the future of the consciously defined entity known as the 'city' (McNeill, 2001a, 2001b).

The issue of narrative coherence is important, primarily because our knowledge of cities is driven by the fact that we can never know them whole, that we can only envision them in parts. Such emblematic spaces tend to be in the spaces of the old downtowns, the 'civic' cores tied to early modern modes of political organization. As Dagger (2000) argues, these spaces are the sites of 'civic memory', 'the recollection of the events, characters and developments that make up the history of one's city or town' (37). With the mayor as first citizen, s/he is ideally placed to reinforce this sense of civic memory: 'When the people and events that formed the city are remembered, the city will be seen not as a curious accident or an incomprehensible jumble, but as something with a story, a past that makes sense of the present' (Dagger, 2000: 38). Thus the mayor can inhabit parts of cities, key sites, and make sense of the city's story through their bodily presence in, or their interpretation of, these spaces.

This preference for making an immediate and direct appearance on the streets of cities is linked to the idea of governmentality as knowing what is to be governed. This involves forms of observation, in the broadest sense – including the collection and processing of statistics, from population growth or decline, to crime rates, to land prices, to traffic numbers. But it can also include the mayor playing the role of 'inspector', arriving unexpectedly to observe how street-cleaning is carried out, or how schools are being run, or how parks are being used. Such site visits may be recycled into campaign speeches and homilies, and may even trigger major policy shifts within city councils.

Conclusions

This chapter has briefly surveyed a range of ways in which mayors fit within urban political theory. It has been suggested that overly structural approaches to politics have no way of explaining the performative dimensions of mayoral governance, which come to the fore at election times, but are also significant in the enactment of difficult and controversial policy choices. To this extent, mayors have agency, both in their ability to construct 'imagined communities' around their cities, but also in the ability to construct the scope of opportunities that the city's executive can pursue. Furthermore, following Low (2007), the chapter has also argued for the importance of analysing the position of the mayor within her or his political party. Even in directly elected mayoral systems, being able to gain the support of a party machine – and control it once in power – is perhaps as challenging a piece of political skill as defeating an electoral opponent. We can certainly nominate the mayor as one of a small number of 'mediating elites' that are capable of bringing together apparently competitive cities into forms of dialogue and co-operation that may aid both (see Allen, 2010).

However, the chapter has also suggested caution about the extent to which mayors – as the most visible expression of state power in any urban polity – should be endowed with excessive agency in political theoretical terms. Certainly, many accounts of urban politics associated with the growth coalition literature have tended to emphasize the importance of private-sector interests in driving urban agendas, even though this may also be overstated in terms of visibility. More significant still is the scalar domination of city councils by both macro-economic forces and central governments, which usually have the power to structure the rules of engagement by which councils and mayors must operate. In terms of power, perhaps it is in the governmentality of everyday life in cities that mayors can exert the most influence, and that may in turn give that mayor sufficient political support to 'jump scales'. However, regardless of how one chooses to define mayoral power (or lack thereof), a grasp of the political geography of the institution, the individual and the represented spatial formation is fundamental.

References

Allen, J. (2010) 'Powerful city networks: more than connections, less than domination and control', *Urban Studies*, 47(13): 2895–2911.

Allen, J. and Cochrane, A. (2010) 'Assemblages of state power: topological shifts in the organization of government and politics', *Antipode*, 42(5): 1071–1089.

Cox, K.R. (1998) 'Spaces of dependence, spaces of engagement and the politics of scale, or: looking for local politics', *Political Geography*, 17(1): 1–23.

Corner, J. (2003) 'Mediated persona and political culture', in Corner, J. and Pels, D. (eds) *Media and the Restyling of Politics: Consumerism, Celebrity and Cynicism*. Sage: London. pp. 67–84.

Dagger, R. (2000) 'Metropolis, memory and citizenship' in Isin, E. (ed.) *Democracy, Citizenship and the Global City*. Routledge: London. pp. 25–47.

Fairclough, Norman (2000) *New Labour, New Language?* Routledge: London.

Fairclough, Norman (2003) *Analysing Discourse: Textual Analysis for Social Research*. Routledge: London.
Giuliani, Rudy W. (2002) *Leadership*. Little & Brown: New York.
Grynbaum, M.M. (2012), 'Mayor warns of the pitfalls in social media', *New York Times*, 21 March.
Joyce, Patrick (2003) *The Rule of Freedom: Liberalism and the Modern City*. Verso: London.
Katz, A. (2010) 'The city that got swapped', *Bloomberg Businessweek*, 26 April–2 May: 100–106.
Lefèbvre, Henri (1991) *The Production of Space*. Blackwell: Oxford.
Logan, John R. and Molotch, Harvey (2007) *Urban Fortunes: The Political Economy of Place*, 20th anniversary edition. University of California Press: Berkeley, CA.
Low, M. (2007) 'Political parties and the city: some thoughts on the low profile of partisan organisations and mobilisation in urban political theory', *Environment and Planning A*, 39(11): 2652–2667.
Massey, D. (2011) 'A counterhegemonic relationality of place', in McCann, E. and Ward, K. (eds) *Mobile Urbanism: Cities and Policymaking in the Global Age*. University of Minnesota Press: Minneapolis, MN. pp. 1–14.
McCann, E. (2011) 'Urban policy mobilities and global circuits of knowledge: toward a research agenda', *Annals of the Association of American Geographers*, 101(1): 107–130.
McFarlane, Colin (2011) *Learning the City: Knowledge and Translocal Assemblage*. Wiley-Blackwell: Oxford.
McNamara, K and McNeill, D. (2012) 'The city personified: the geopolitical narratives of Rudy Giuliani', *Communication and Critical/Cultural Studies*, 9(3): 259–278.
McNeill, D. (2001a) 'Barcelona as imagined community: Pasqual Maragall's spaces of engagement', *Transactions of the Institute of British Geographers*, 26(3): 340–352.
McNeill, D. (2001b) 'Embodying a Europe of the Cities: geographies of mayoral leadership', *Area*. 33(4): 353–359.
Polner, R. (2005) 'Introduction', in Polner, R. (ed.) *America's Mayor: The Hidden History of Rudy Giuliani's New York*. Soft Skull Press: New York. pp. xix–xxxv.
Weber, R. (2010) 'Selling city futures: the financialization of urban redevelopment policy', *Economic Geography*, 86(3): 251–274.

7

MAKING URBAN POLITICS GO AWAY: THE ROLE OF LEGALLY MANDATED PLANNING PROCESSES IN OCCLUDING CITY POWER

John Carr

Introduction: The mundane frustrations of public input-based urban planning

A political decision is made in city hall and a planning process is ordered. Notices are mailed and signs are posted on city property. Public meetings are held. Planners show PowerPoint presentations to irate neighbours who raise their voices and submit impassioned letters and e-mails in opposition. A draft plan is prepared and another round of meetings is held. A smaller group of irate neighbours shows up to voice their frustration with the draft. A slightly modified plan is presented to the Mayor and City Council where it passes handily amid statements of the robust public input and democratic process that led to its passage. And with the passage of the plan, the original decision transforms into policy.

Throughout the US this scenario is likely a familiar by-product of legally mandated participatory planning processes that work to render political decisions apparently apolitical. At first glance, the interplay between planners and citizens lacks many of the hallmarks we commonly associate with urban politics. Intentionally designed to exclude elected officials and high-level administrators, far removed from campaign promises and press conferences, often conducted in community centres and school gymnasiums, the realm of participatory planning often appears to be a backwater that is far removed from the 'big show' of city hall politics. Yet if we properly understand urban politics as encompassing all range of decisions about how the varied costs and benefits of urban form are distributed across society (Davies and Imbrioscio, 2009; Fainstein, 2005; Harvey, 1973; Purcell, 2003), then planning becomes an essential site for interrogating how these political decisions get made (Sandercock, 1998).

To begin to outline both how and why elites filter their political decisions through participatory planning processes, this chapter offers the case study of Seattle, Washington's efforts to accommodate a burgeoning skatepark advocacy movement. This study suggests that participatory planning enables political elites – including elected officials, effective advocacy organizations, businesses, and well-organized neighbourhood groups – to reframe their policy preferences as the product of ground-level democratic processes, and open, fair, and impartial technocratic decision-making. Indeed, it is this very tendency that has ensured the longevity of participatory planning, notwithstanding the cynicism with which citizen participants (Buttny and Cohen, 2007; Herbert, 2006), planning professionals (Brooks, 2002; McComas, 2001), and elected officials (Culver and Howe, 2004) view it.

In Seattle, as in many other US cities, certain governmental decisions are required by law to be funnelled through participatory planning processes. And while there is a rich and varied literature arising from critical legal studies that traces the way the ostensibly neutral legal system is designed to legitimate inherently political matters by transforming them into apparently a-political judicial determinations (Gordon, 1984; Kennedy, 1998; MacPherson, 1962; Unger, 1976), relatively little attention has been paid to the similar functions that planning performs – notwithstanding powerful work interrogating the class and power biases of many traditional approaches to planning (Friedmann, 1987; Harvey, 1978; Sandercock, 1998). Like other legally mandated and formally structured mechanisms, participatory planning must be understood as a tool for selectively making politics 'go away'.

Understanding Law and Planning as Tools for Occluding Power Politics

Contemporary practices of urban planning, with their emphasis on process, procedure, and expert framing are both analogous to, and an extension of the ways that the legal system works to veil the political nature of much governmental decision-making. There is a robust literature arising from Critical Legal Studies ('CLS') that argues that much of the legal system's function is to remove profound questions about social justice and inequality from the political arena (see Box 7.1). Under this perspective, '[l]aw is simply politics dressed in different garb; it neither operates in a historical vacuum nor does it exist independently of ideological struggles in society' (Hutchinson and Monahan, 1984: 206). Thus, the aggressive policing of homeless populations (Mitchell, 2003), or the use of private property rights to force poor renters out of a gentrifying neighbourhood (Blomley, 2004) cannot simply be considered the result of impartial policing

> ### BOX 7.1 CRITICAL LEGAL STUDIES AND URBAN GEOGRAPHY
>
> Critical Legal Studies (CLS) and its various critical offshoots (including approaches to understanding the law informed by feminism, critical race theory and ecofeminism) all begin from the premise that the law should be understood not by what it purports to do, but by the practical results it produces. For CLS influenced scholars, even those aspects of the legal system that we celebrate as ensuring fairness and equality under the law – including uniform legal and procedural rules, the right to an impartial jury of one's peers, and the elimination of legal distinctions based on race, religion and gender – should be evaluated to see if they actually produce results that are fair and equal in practice. Thus, some of the earliest foundational studies in CLS – including Kennedy's (1979) work on Blackstone's 'Commentaries' (which served as the foundation of US common law) and Unger's (1976) investigation into the intertwined Enlightenment roots of modern legal and political forms – trace the ways that modern legal systems developed to allow the perpetuation of pre-existing patterns of social inequality under the rubric of universally applicable laws and equal rights (Gordon, 1984).
>
> An almost dizzying array of work critical of the political functions of the legal system has followed, with a substantial portion exploring the ways the law simultaneously animates urban geographies, while occluding and naturalizing its own influence. This research has traced such diverse topics as the translation of social understandings of place into geographically differentiated approaches to the law (Engle, 1984), the law's role in defining and enforcing racial boundaries in cities (Delaney, 1998), the power of private property to shape urban geographies (Blomley, 2004, 2007), the emergence of innovative legal tools to exclude 'undesirable' populations from cities (Beckett and Herbert, 2009), and the tendency of the law to 'sort' bodies in the city along normative conceptions of citizenship (Carr et al., 2010).
>
> While emerging from a variety of disciplines and perspectives, these works all share a profound sensitivity to the tendency of law to represent itself as a 'level playing field' for the neutral, dispassionate and a-political resolution of urban conflicts, while simultaneously enabling the predetermination of those conflicts along a variety of economic, political and social axes.

practices, or of neutral legal logics impartially administered by the courts. Rather, they must be understood to be just as political as municipal budgets and the composition of City Council. Indeed, CLS suggest that the work done by legal actors constitutes a particularly insidious form of politics, one that masks itself behind the promise of formal equality under the law, and highly structured yet often impenetrable systems of procedures and rules.

There are at least two reasons to analyse contemporary urban planning within a CLS context. First, planning practices are commonly integrated with political processes by legal mandate. Laws requiring direct citizen involvement with urban planning have increased at the local and federal level since the 1960s when a public backlash against urban renewal and slum clearance programmes opened the door to citizen involvement (Brooks, 2002). In response to this outcry, new regulations established 'Citizen Conservation Councils' to advise on federal urban programmes (Gil and Lucchesi, 1979), with most local governments following suit (McComas, 2001). What has resulted is a now standardized set of steps for soliciting citizen input in response to contemplated governmental actions, without any concomitant mechanism for uniformly translating that input into policy. These include 'public hearings, written public comments on proposed projects, as in environmental review, and the use of citizen-based commissions, such as planning and zoning commissions, and boards of directors for public agencies with quasi judicial, and/or quasi legislative power along with advisory committees and task forces' (Innes and Booher, 2004: 423).

Second, planning should be understood as part of the juridical mechanism for distributing the costs and benefits of social membership in cities due to the clear parallels between traditional legal adjudication and public input-based planning processes. Just as the judicial system has long positioned itself as an a-political agency for the dispassionate discovery of truth under a system of impartial procedural rules, so too has planning been framed as a realm whose processes are ostensibly designed to remove the taint of traditional power politics from urban decision-making. Both legal and planning procedures are characterized by a series of tightly focused hearings and appeals in which experts (lawyers, planners) work to influence outcomes by translating lay-arguments into technical terms and provide official interpretations, and whose merits are determined by an authority (judge, jury, planning board, task force, etc.) that may gather information but provides no feedback to participants until a final decision is rendered. Both realms are characterized by tightly enforced rules regarding advance notice of procedural events and the manner in which information may be presented, as well as a complex system of guidelines, requirements, and precedents that are largely inaccessible to the average citizen.

Working together, these shared characteristics enable both urban planning and the judicial system to legitimate and naturalize political decisions about who will and will not reap the rewards of social membership, reframing them as products of neutral and impartial processes. The types of projects that are currently enabled by participatory planning – including the creation of public space, the destruction and redevelopment of poorer residential areas, land-use rules, and the location of transportation and other infrastructure (Slater, 1984) – all involve substantial decisions about where and for whom the costs and benefits of social membership will accrue.

By routing these decisions through participatory planning, however, their profoundly distributive (and thus political) nature is largely occluded by a process that both obscures and enables the workings of power in the city. Public meetings, hearings, formal rules, and the proliferation of input providing bodies (advisory boards, subcommittees, neighbourhood associations) all serve to diffuse the moments and means by which an elite mandate or political compromise among powerful interests is incrementally transformed into policy, while clothing that mandate in democratic garb. And by spreading this change across a process that is as prolonged and intensive as the typical participatory planning exercise, the very elites spearheading a given transformation in the urban environment – whether mayor, influential council members, lobbyists, activists, or businesses – are able to selectively cloak their role in doing so.

Using Planning to Legitimate Seattle's Public Skatepark Politics

The tendency of legally mandated planning processes to render underlying political decisions invisible is illustrated by Seattle's recent experience accommodating an insurgent public skatepark movement. Here, enacting policy through public input-based processes allowed decision-makers to both provide controversial choices with a veneer of participatory democracy, and to control the extent and nature of that cover. Where the public input elicited by such processes proved a policy to be a political 'winner', elites could claim that victory as their own by claiming sponsorship of the underlying planning exercise. In contrast, where a decision was politically threatening, channelling that decision through planning allowed elites to attribute responsibility for their preferred outcomes to a 'fair' process, an 'expert' planner, and/or the will of the people as manifested in public comment.

Between 2003 and 2007, I was involved with approximately ten different public skatepark projects in and around Seattle as a political lobbyist, an organizer and activist, a member and, subsequently, the chairperson of the Parks department's recently created citizen advisory Skatepark Advisory Committee

(SPAC), a member of the 'Taskforce' supervising a city-wide master planning process for a system of skateparks, and as an informal advisor to the organizers of individual projects. My involvement followed the City's announcement at the end of 2003 that one of only two public skate facilities in Seattle would be demolished to make way for the passive use Ballard Commons park. This decision sparked the creation of a substantial public skatepark movement revolving around such organizations as the Puget Sound Skate Park Association (PSSA) and Parents for Skate Parks (PFS). Not only did Seattle's skatepark movement succeed in saving a place for skateboarders in the new Ballard Commons Park, but the political capital gained through that struggle was also then successfully transferred into a succession of new public skatepark projects.

Figure 7.1 Ballard Commons Park with the New Ballard Skatepark (source: author)

From Policy to Planning

Throughout my time in Seattle, ostensibly democratic and participatory processes were used to ratify informal policy decisions by urban elites. And while this criticism has been levelled for some time (Checkoway, 1981; Dennis, 2006), evidence from the skatepark advocacy movement suggests that it is the flexibility with which participatory planning can either obscure or reveal the workings of informal power politics that at least partially explains its ongoing popularity and legal mandate.

The law in Seattle is typical of the national trend requiring citizen-input-based planning whenever city governments contemplate substantial changes to the urban fabric. Washington State adopted a 'Growth Management Act' (Revised Code of Washington (RCW) 36.70A) in 1990, that requires now familiar forms of notice (RCW 36.70A.035) and public participation, including 'broad dissemination of proposals and alternatives, opportunity for written comments, public meetings after effective notice, provision for open discussion ... and consideration of and response to public comments' (RCW 36.70A.140) for all 'coordinated land use polic[ies adopted by] the governing body of a county or city' (RCW 36.70A.030(4)). Likewise, Seattle's Municipal Code creates a variety of opportunities for public input and process, resulting in a plethora of official and semi-official agencies, councils, committees and groups that have contributed to a citywide 'reputation – or perhaps even notoriety – for public process' (Mattern, 2003: 5). For example, the Seattle code creates an 'Associated Recreation Council' to represent neighbourhood 'community advisory committees' in dealing with parks issues throughout the city (Seattle Municipal Code (SMC) 18.04.010), and a volunteer advisory 'Board of Park Commissioners' to make recommendations regarding planning and policy (SMC 3.26.030(B)). In turn the Parks Board and the Superintendent are required to hold public meetings periodically to review Parks programmes and objectives (SMC 3.26.030(E), SMC 3.26.040(K)). These State and City rules are buttressed by an internal Parks policy that has enshrined public participatory planning. When dealing with recreational spaces Parks must solicit, 'direct citizen involvement, participation and public input' both in planning for future needs and once funding is secured for specific projects (Seattle Parks and Recreation, 1999: 4).

Notwithstanding these ostensibly well-intentioned mandates to render urban decision-making more democratic, Seattle's political elites have treated planning procedures as a means to achieve their own ends when dealing with public skateparks, rather than as tools for actually guiding policy. This tendency was perhaps most graphically demonstrated by the project that gave rise to Seattle's skatepark movement, the Ballard Bowl. Initially created as a DIY ('do it yourself') project on the site of an abandoned grocery store that had been purchased for future Parks development, the original Ballard skatepark became the focal point for a loosely affiliated group of adult skaters, many of them in the software and information industries. When the City announced in 2003 that the skate features would be razed to make way for a passive use park, these skaters formed the core of the nascent skatepark movement. And while the effort to save the Bowl was initially a minor local cause, organizers quickly transformed the movement into an effective and, I would argue, elite political force through three tactics. First, activists reframed the Ballard controversy as part of a broader citywide failure to meet the needs of young people, while offering the construction of more skateparks as the answer to that problem. Second, they conducted a successful media campaign heightening the profile of the cause through a series of events including a skate jam/fundraiser, a 'thousand skater march' from the

Seattle Space Needle to City Hall, and a 'wake'/benefit concert for the Ballard Bowl. Third, a core group of activists of which I was a member, successfully lobbied key Council members to support the Ballard facility and to back the cause of public skateparks as a youth amenity. With the political and media support of these elected officials, advocates managed to reverse city policy, saving a space for skateboarding in the soon to be developed Ballard Commons park.

And while this was an unquestionable victory for the new advocacy movement, it also revealed some harsh truths about elites' instrumental attitudes towards public input, as the subsequent decision-making around Ballard's skatepark was animated solely by pressure from political players operating outside of, and often in contravention of, prior participatory planning exercises. Planning for the Ballard Commons had begun in 1994 with the creation of a neighbourhood-based 'Ballard Open-Space Committee,' which ultimately led – through traditional public input processes – to the 1998 'Crown Hill/Ballard Neighborhood Plan'. The Plan recommended the creation of a passive use park as an anchor for a new Civic Center within the commercial core of the traditionally Scandinavian, maritime Ballard neighbourhood. Another planning process conducted in 2000 involved intensive neighbour and stakeholder input, generating a set of design elements and goals for the new park. Thus, when the Mayor and Parks Department announced that, notwithstanding skateboarders' late appearance and lack of involvement in prior participatory exercises, their demands would override the stated preferences of prior planning participants for a quiet, passive-use park, many were outraged including the Ballard Chamber of Commerce, several local business owners, and a number of vocal neighbourhood activists who had been involved in the ongoing participatory planning for the Ballard Commons. Notwithstanding their own lobbying and media efforts, these skatepark opponents found themselves unable to overcome the support of a Mayor and City Council that had been persuaded to jump aboard the skatepark bandwagon as a high-profile, timely, and youth-oriented cause (Carr, 2012). Likewise, skaters found themselves too late to overcome the backroom influence of a neighbouring property owner that persuaded the Mayor's office to 'move' (that is, demolish and rebuild) the skatepark fifty feet away from its original location, eliminating the youth-friendly 'street' features that had comprised approximately two-thirds of its area. Thus, by overriding a long-standing, neighbourhood-wide participatory planning project in an effort to accommodate effective private lobbying by two special interests, City government signalled that public input matters so long as it does not conflict with elected officials' priorities.

Seattle's efforts in 2006 and 2007 to create a citywide master plan for future skateparks not only demonstrated a similar lack of deference to participatory planning processes, but it also illustrated some of the ways that political elites are able to use public-input to 'depoliticize' policy mandates. The master planning exercise was the product of a de facto partnership between the Mayor's office, City Council, and relatively privileged and empowered skatepark advocates like me. As mentioned above, that informal alliance began with a series of meetings between individual council members and representatives of the PSSA and PFS to discuss the Ballard skatepark issue. During those meetings, advocates successfully recruited key Council members – including the entire Parks subcommittee – to support a broader pro-skatepark initiative. The core of this initiative was a master plan of future locations intended to avoid the type of site-by-site NIMBY ('not in my backyard') resistance that we had encountered with Ballard and two other pending projects.

The resulting political partnership dictated the results of the ensuing masterplanning exercise, which was largely treated as a means, rather than an end in itself. From its inception, the Task Force upon which I sat, Parks staffers, and the planning firm all envisioned their jobs as identifying good sites to facilitate the building of new facilities, rather than determining public demand or need. Our shared vision was clearly manifested in the formal planning firm presentations at each public meeting, which largely served as sales pitches for public skateparks. Each presentation offered data on skateboarding's expanding popularity while attempting to rebut concerns associating skateparks with danger, crime, and disorder. In fact, a number of comments objected to the use of public meetings to promote skateparks rather than engendering meaningful dialogue. As one e-mailed comment to the Parks department argued: '[t]he city may have valid reasons for promoting skateparks, but those reasons are not evident from the data provided ... A presentation on how great skateboarding is, even if completely accurate, doesn't help to identify appropriate siting criteria ... Stop trying to "educate" the general public and start listening.'

Notwithstanding our promotional efforts, the master plan drew substantial public opposition, with the Parks department receiving over two hundred e-mails, approximately twenty letters and written comment cards, and dozens of phone calls in opposition to individual sites and/or the broader project of creating a network of potential skatepark locations – in contrast to approximately ten unconditionally supportive letters and/or e-mails. At each public meeting, public input was dominated by anti-skatepark comments at a rate of over five to one, in spite of our efforts to ensure the presence of a vocal pro-skatepark constituency. Objections ranged from concerns with loss of green space, to negative impacts on property values, to worries about danger and criminality associated with teen skaters. As one self-identified business owner argued via e-mail in opposition to a proposed site in his neighbourhood, '[s]kateboarding is an excessively noisy activity which can easily result in serious injury and often correlates with destruction of property ... I am fed up with the Parks Department shortchanging its responsibility for environmental stewardship in favour of funds being allocated to babysitting youth whose parents have abdicated their responsibility.'

Even so, most opponents found themselves either too late to shut down the master plan, or forced to channel their resistance through a series of technically oriented and often impenetrable meetings of the Task Force, the Department of Parks and Recreation, and/or City Council. And most objections to specific proposed locations were largely discounted by Parks staffers, Task Force Members, and planning professionals alike as being either founded in negative stereotypes about skateboarders, capable of redress through sound design and planning practice, or simply misguided. The only exceptions to this tendency were those neighbourhoods that had the political clout to have nearby facilities simply removed from (or never included in) the draft plan (Carr et al., 2010). Perhaps unsurprisingly, the final master plan prepared by the Task Force and duly approved by City Council on 31 January 2007 included every proposed skate park location that had been pre-approved by the Parks department, save those that had been excluded for concerns about political fallout from elite neighbourhoods or the mayor's staff.

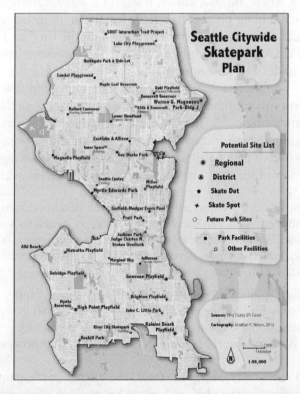

Figure 7.2 Seattle's 2007 Master Plan for Skateparks (source: author)

Taken together, this process hardly represented the type of ground-level democracy-enhancing public participation that 'increases social capital and empowers citizens as they seek a stronger voice in decisions that affect their lives' (Laurian, 2004: 53) as participatory planning is intended to do. Rather, from the perspective of those like me who were invested in 'getting skateparks built', public input served as a box to be checked – more or less successfully – along the way. As one of my fellow advocates noted, tacitly acknowledging the tendency of planning to serve policies already decided upon by city elites, 'I hope they [Parks] just use their bullshit manufactured process to just railroad this thing through . . . We are so used to them doing what they want and then processing it, but this time I hope they [actually] do [just that].'

The important lesson from Seattle is not simply that elites can 'stage manage' (Maginn, 2007) ostensibly democratic planning processes to achieve desired political ends. Rather, planning processes served as a type of camouflage that elected officials used to obscure their decision to support skateparks when those projects encountered resistance, and then claimed credit once that resistance was overcome. Tellingly, in reviewing hours of public testimony and hundreds of pages of written public comments to various skatepark projects, not once did I ever observe or hear a complaint about the mayor, individual City Councillors, or even the Council as a whole. Instead, citizen frustration was invariably aimed at those entities that the planning process had rendered 'at hand' including planners, the Parks department and individual staffers, as well as 'the City' as a whole.

Citizen participants consistently addressed their concerns to planners and Parks staffers under the assumption that it was these functionaries who ultimately wielded the discretionary power to locate and build new facilities. And while the tendency to conflate decision-making authority with the low-level city functionaries attending public meetings is understandable, it was an impression that was consistently reinforced by those very functionaries. The following exchange from one of the question and answer sessions following the 2 October 2006 presentation of the draft master plan typifies this dynamic:

> Citizen Participant: I come from the Genesee neighbourhood, and we want to support you but at the same time we want to support our green space. Will you allocate money to planting more trees or do some other form of reparations to the park to help reduce the impact of the skate facility being built?
>
> Parks Staffer: Right now, the Parks department has no money for any of this. They only have money for the planning process. So in the design of a future site, if the community decides that trees are very important, then that is something that would get folded into that process.

As with the vast majority of public input, the citizen participant addresses the planners and agency staffers before her as the 'you' who is presumed to be the decision-maker. And, in turn, the responding Parks staffer accepts and confirms this presumption, readily shifting the locus of authority for public space decisions between the Parks department (a purportedly non-political bureaucratic agency), and some future planning exercise that will ostensibly be driven solely by the public's desires.

In the rare instance that City Council was mentioned at all, its involvement was typically framed as enabling citizen-driven planning. For example, during the 2 October 2006 meeting, the following exchange presented City Council as subservient to public input in funding skateparks, notwithstanding its role in driving their creation:

> Citizen Participant: What kind of commitment do you have from city council, like we're going to do all this planning and the design, so what's their promise?
>
> Planner: Well, I think the important thing to realize is that they recognize that skateboarding is such a fast-growing sport and they do want to provide facilities. Otherwise they wouldn't have given us money to do this plan in the first place. So the idea is hopefully we'll have a really good process, a really good open dialogue, really good feedback, we want to address everybody's concerns, and then so hopefully when this all gets presented to city council they will see that we've done a lot of hard work as a community and we've gotten good feedback and the Taskforce has recommended sites that everybody is going to be in agreement about. So hopefully that will inspire them to earmark some money for the project.

Throughout, planners and Parks staffers served as more than the public 'face' for Seattle's experiment with public skateparks. Rather they embodied the state in such a way that it was 'their' proposed project, process, or bureaucratic agency that was ultimately held accountable for designing and allocating the finite and contested municipal good of public space. And by allowing these relatively low-level functionaries to bear the brunt of public opposition and frustration, the essential role of such elites as the Council and the Mayor in 'getting skateparks built' was effectively occluded behind the quasi-legal formalism of the planning process itself.

From Planning Back to Politics

It is not enough, however, for planning to simply help elites obscure their involvement with urban change in the face of public resistance. If the politics of public space truly do disappear, then so does the political capital that elected officials hope to garner by backing planning-enabled projects. The beauty of participatory planning is that it serves as both sword and shield for such elites, allowing them to channel negative feedback into politically inert staff-level planners and functionaries, while simultaneously demonstrating their leadership and commitment to democracy. In Seattle, once skateparks become 'facts on the ground', the shroud of bureaucratic process is typically shrugged off with elected officials stepping up to both claim credit before their constituencies and to pay respect to the legitimating democratic processes that led to their policy victories.

This was dramatically illustrated at the Ballard Commons park. By the time it opened, most major constituencies had been substantially frustrated by its generative process including the proponents of a passive-use Commons, the skaters who had lost two-thirds of the footprint and the signature bowl feature of the original skatepark, and the developers of the adjacent multi-use development who feared the skaters would scare away buyers.

Figure 7.3 Opening day for Ballard Commons Park, with the skatepark in the background
(source: author)

Even so, Seattle's elected officials quickly managed to consign such disappointments to the past. The opening ceremonies took up the better part of an unseasonably balmy March day in 2006, with the Mayor, the Superintendent of Parks, and the majority of City Council showing up to celebrate the City of Seattle's ability to create new public space for its citizens. And as has been typical for all of Seattle's successful skatepark projects, the rhetoric that day celebrated the new facility, the participatory processes that led to it, and the politicians who claimed credit for both. Mayor Nickels' opening speech rigorously followed this formula:

> We have people who are here today who have been involved in the process of creating this park for a decade ... So congratulations to the people of Ballard who have been patient while this has been taking shape and particularly to those who came up with this vision and saw it through to the community plan ... It is because of the work of the council that the pro-parks levy was on the ballot and we're enjoying and celebrating days like this all over the city.

Similarly, one of the City Council members who had had been a particularly active supporter likewise celebrated both the participatory planning exercise and his own leadership in that process:

> I want to thank all of the people who have stood in there for the past ten years and have struggled together to come up with this fine park. I want to lastly say that council member Drago and I were some of the few people that stood with the skateboarders to make sure that they had a place in this park as well.

More than an opportunity for elected officials to claim a democratically legitimated political victory, the opening ceremony also served as a moment for even previously aggrieved constituencies to step 'in line' behind those politicians, further occluding the political compromises, exercises of executive fiat, and anti-democratic moments that had characterized its construction. For example, a prominent neighbourhood advocate who had strenuously objected to the skaters' eleventh-hour claim to the Commons made a brief speech praising a variety of individuals and organizations who had been involved in almost ten years of public input, celebrating the ways that 'neighbours like you were involved in each and every step' with 'the skateboard community, nearby neighbours, the chamber of commerce, [and] the Ballard open space committee ... work[ing] together to make this place'. Likewise, I spoke on behalf of the SPAC and – eager not to alienate the political elites whose favour my movement would continue to depend upon – framed the Ballard compromise as a win–win result for skaters and the Ballard neighbourhood, while explicitly thanking the Mayor, the Council, and Parks for their patronage.

As a whole, the Ballard opening ceremonies were typical of the ways Seattle's often contentious, non-democratic, and elite-driven politics of public space have been reframed and legitimized as citizen-driven, elite enabled, and inherently consensus based. And because participatory planning enables dissenting voices to be diffused through depoliticized local meetings and redirected at low-level functionaries, the organizers of the Ballard opening could freely proffer such framings without fear of dissenting voices disrupting the proceedings.

Conclusion: The Subtlety and Invisibility of Planning as a Manifestation of Legal Formality

I offer the foregoing analysis and case study not to disown my work as a skatepark advocate, but rather to offer a critique of the skewed and anti-democratic features of a common planning process we found ourselves working within. And given the substantial literature criticizing participatory planning (Anderson et al., 2003; Ataov, 2007; Bens, 1994; Culver and Howe, 2004; Innes and Booher, 2004), it is clear that many who deal with city politics are likely familiar with the sometimes grinding political, economic, and scheduling costs that come from routing urban decisions through these processes. In Seattle, as elsewhere, the inherent problems with such processes are typically seen as a constant but low-level inconvenience rather than as a source of injustice – ranking somewhere between inefficient bureaucracy and sparse municipal budgets in the scale of ubiquitous and permanent impairments to getting things done in the city. And it is this sense of a-political inevitability that makes participatory planning so problematic and so potentially powerful as a tool for obscuring the informal workings of elite political manoeuvring.

While participatory planning processes frequently function both through legal mandate and in the manner of legal processes, I would argue that they have an even greater capacity to make politics 'go away' than conventional judicial mechanisms due to their low visibility. With US Federal trial courts publicly wading into such intensely political waters as health-care reform, state immigration laws, and the right to same-sex marriage, it is unlikely that most Americans are unaware of the potential for the court system to serve as a tool for politics 'by other means'.

In contrast, legally mandated participatory planning processes still remain potently effective in obscuring the links between elected officials, municipal policy, and the urban environment precisely because they have maintained the appearance of being removed from broader circles of city power, thus achieving 'an almost sacred quality' that has rendered them immune to frequently levelled criticisms (Innes and Booher, 2004). And one must assume that this sacrosanct quality is due, in no small part, to the political establishment's understanding that even the most cutting-edge and earnestly implemented efforts to render participatory planning democratic, representative, and inclusive can still be manipulated to produce the desired results of elites (Flyvbjerg, 1998; Maginn, 2007). Experience from Seattle reinforces this sense, given that – in the case of the Parks department – participatory planning exists largely as a matter of executive fiat. Parks' policies requiring public input have been adopted under the discretionary authority of the Parks Superintendent (SMC 3.26.040(C)). In turn, the Superintendent serves at the pleasure of the Mayor, subject to confirmation by a majority of City Council (Seattle City Charter, Article XI, Section 1). At a practical level, this means that participatory planning for parks likewise exists – and could simply be eliminated – at the pleasure of the mayor, with the tacit consent of City Council.

Thus, to the extent that Seattle is typical of broader tendencies, I would suggest that long-standing efforts among academics and practitioners to 'fix' planning are misguided. Notwithstanding the almost universal adoption of participatory planning processes by cities around the US, there is not now and has never been a consensus on how elected officials and staffers should treat the 'ground level' and ostensibly democratic input solicited by those processes. Rather, questions around the nature, proper use, and value of public input are continually debated by academics and planners, while being resolved on an ad hoc basis by policymakers (Innes and Booher, 2004); as was the case in Seattle. It is this very indeterminacy in the relationship between public input and policy that opens the door to the use of formalistic planning processes to occlude the workings of elite level politics. Accordingly, I would argue that, at the very least, a critical and comprehensive re-evaluation of what participatory planning is and does in the real world is long overdue. And if experience from sites as diverse as Seattle, Washington, Aalborg, Denmark (Flyvbjerg, 1998), and Perth, Australia (Maginn, 2007) is representative of broader trends in urban planning, then that project as a whole must be reconsidered and likely abandoned.

References

Anderson, R., Cissna, K. N. and Clune, M. K. (2003) 'The Rhetoric of Public Dialogue', *Communication Research Trends*, 22(1): 1–33.

Ataov, A. (2007) 'Democracy to Become Reality: Participatory planning through action research', *Habitat International*, 31: 333–334.

Beckett, Katherine and Herbert, Steve (2009) *Banished: The New Social Control in Urban America*. Oxford University Press: New York.

Bens, C. K. (1994) 'Effective Citizen Involvement: How to make it happen', *National Civic Review*, 83(1): 32–39.

Blomley, N. (2004) *Unsettling the City: Urban Land and the Politics of Property*. Routledge: New York.

Blomley, N. (2007) 'How to Turn a Beggar into a Bus Stop: Law, traffic and the "function of the place"', *Urban Studies*, 44(9): 1697–1712.

Brooks, Michael P. (2002) *Planning Theory for Practitioners*. Planners Press: Chicago.
Buttny, R. and Cohen, J. R. (2007) 'Drawing on the Words of Others at Public Hearings: Zoning, Wal-Mart, and the hreat to the aquifer', *Language in Society*, 36: 735–756.
Carr, J. (2012) 'Public Input/Elite Privilege: The use of participatory planning to reinforce urban geographies of power in Seattle', *Urban Geography*, 33(3): 420–441.
Carr, J., Brown, E. and Herbert, S. (2010) 'Inclusion under the law, exclusion from the city: The regulation of bodies and places in Seattle', *Environment & Planning A*, 41(8): 1962–1978.
Checkoway, B. (1981) 'The Politics of Public Hearings', *Journal of Applied Behavioral Science*, 17(4): 566–582.
Culver, K. and Howe, P. (2004) 'Calling all Citizens: The challenges of public consultation', *Canadian Public Administration*, 47(1): 52–76.
Davies, J. S. and Imbrioscio, D. L. (2009) 'Introduction: Urban politics in the twenty-first century', in Davies, J. S. and D. L. Imbrioscio (eds) *Theories of Urban Politics*, Second Edition. Sage: Thousand Oaks. pp. 1–14.
Delaney, D. (1998) *Race, Place and the Law, 1836–1948*. University of Texas Press: Austin, TX.
Dennis, S. F. J. (2006) 'Prospects for Qualitative GIS at the Intersection of Youth Development and Participatory Urban Planning', *Environment & Planning A*, 38(11): 2039–2054.
Engle, D. M. (1984) 'The Oven Bird's Song: Insiders, outsiders, and personal injuries in an American community', *Law and Society Review*, 18(4): 551–582.
Fainstein, S. S. (2005) 'Planning Theory and the City', *Journal of Planning Education and Research*, 25(2): 121–130.
Flyvbjerg, Bent (1998) *Rationality and Power: Democracy in Practice*. University of Chicago Press: Chicago.
Friedmann, John (1987) *Planning in the Public Domain: From Knowledge to Action*. Princeton University Press: Princeton, NJ.
Gil, E. and Lucchesi, E. (1979) 'Citizen Participation in Planning', in So, F., Stollman, I., Beal, F. and Arnold, D.S. (eds) *The Practice of Local Government Planning*. ICMA: Washington, DC. pp. 552–575.
Gordon, R. W. (1984) 'Critical Legal Histories', *Stanford Law Review*, 57: 57–126.
Harvey, D. (1973) *Social Justice and the City*. Johns Hopkins Press: Baltimore, MD.
Harvey, D. (1978) 'On Planning the Ideology of Planning', in Burchell, R. and Sternlieb, G. (eds) *Planning Theory in the 1980s: A Search for Future Directions*. Center for Urban Policy Research: New Brunswick, NJ.
Herbert, S. (2006) *Citizens, Cops, and Power: Recognizing the Limits of Community*. University of Chicago Press: Chicago.

Hutchinson, A. C. and Monahan, P. J. (1984) 'Law, Politics, and the Critical Legal Scholars: The unfolding drama of American legal thought', *Stanford Law Review*, 36: 199–246.

Innes, J. and Booher, D. E. (2004) 'Reframing Public Participation: Strategies for the 21st century', *Planning Theory & Practice*, 5(4): 419–436.

Kennedy, D. (1979) 'The Structure of Blackstone's Commentaries', *Buffalo Law Review*, 28: 205–382.

Kennedy, D. (1998) *A Critique of Adjudication [fin de siècle]*. Harvard University Press: Cambridge, MA.

Laurian, L. (2004) 'Public Participation in Environmental Decision Making', *Journal of the American Planning Association*, 70(1): 53–65.

MacPherson, Crawford B. (1962) *The Political Theory of Possessive Individualism: Hobbes to Locke*. Oxford University Press: Oxford.

Maginn, P. J. (2007) 'Deliberative Democracy or Discursively Biased? Perth's dialogue with the City Initiative', *Space & Polity*, 11(3): 331–352.

Mattern, S. (2003) 'Just How Public Is the Seattle Public Library? Publicity, posturing, and politics in public design', *Journal of Architectural Education*, 57(1): 5–18.

McComas, K. A. (2001) 'Theory and Practice of Public Meetings', *Communication Theory*, 11(1): 36–55.

Mitchell, Don (2003) *The Right to the City: Social Justice and the Fight for Public Space*. Guilford Press: New York.

Purcell, M. (2003) 'Excavating Lefèbvre: The right to the city and its urban politics of the inhabitant', *GeoJournal*, 58(2–3): 99–108.

Sandercock, L. (1998) 'The Death of Modernist Planning: Radical praxis for a postmodern age', in Douglass, M. and Friedmann, J. (eds) *Cities for Citizens: Planning and the Rise of Civil Society in a Global Age*. John Wiley & Sons: Chichester. pp. 163–184.

Seattle Parks and Recreation (1999) *Public Involvement Policy for Parks Planning Processes and for Proposals to Acquire Property, Initiate Funded Capital Projects, or Make Changes to a Park or Facility*. Retrieved 13 August 2010. from www.cityofseattle.net/parks/Publications/Policy/PIP.pdf.

Slater, David C. (1984) *Management of Local Planning*. International City County Management Association: Washington, DC.

Unger, Roberto M. (1976) *Law in Modern Society: Toward a Criticism of Social Theory*. Free Press: New York.

SECTION 3
CITY AS COMMUNITY

INTRODUCTION TO CITY AS COMMUNITY

Deborah Martin and Mark Davidson

community [com-mu-ni-ty] ~ a social group of any size whose members reside in a physical or virtual locality, share government and/or often have a common cultural and historical heritage

In the final collection of chapters in the text, we explore the city as community. We use the term 'community' broadly here, since each author will offer a slightly different interpretation of community. If we take 'community' to mean some kind of social grouping, we can identify all types of community associations that any one individual might have in the city. For example, you might be a member of the female, working-class, ethnic minority who votes for green candidates in elections. Our entry point of 'community' there examines the city has a space that has, and indeed constructs, certain types of communities.

Marx famously described capitalist societies that were divided into social classes as alienated (*entfremdung*). Social relations between human beings had become structured in and through the capitalist economy. This alienation was three-fold, involving work being alienated from the products of labour, the act of labour itself and, lastly, workers being alienated from other workers. This type of commentary should not be unfamiliar by now. If we look back to the early urban sociologists, we find descriptions of the ways in which close social bonds and associations had been replaced by weak and strategic relations that were structured through the market place. So what does constitute community in this context?

Since the 1980s urban geographers, amongst others, have been much more concerned with this question. Feminist, postmodern and post-structuralist theory all helped bring attention to the various forms of identity and community that had remained overlooked by those focused on politico-economic approaches to the city. Rather than simply seeing alienated individuals and class divisions, the urban literature began to reflect a concern with gender, race, ethnicity, sexuality and so on. What these literatures have done is make urban scholarship much more cognizant of the numerous modes of community that go into making up the urban populace. In doing so we are now much more aware of the contestations and struggles that various social groups have undertaken in the city. Many of these struggles have taken place either outside or at an arm's length from the state. As such many traditional state-centric approaches to urban politics might miss them. The question of whether to focus upon urban politics as a state-based phenomenon is a contentious one.

If we follow the theory of politics that Jacques Rancière (1999) developed, then we find an emphasis upon 'the part of those who have no part'. These social groups, who are only defined by their unequal status in society, are by definition disconnected from institutional politics. The state is the site of policing and, as such, must be devoid of latent politics unless it is fully realizing its democratic status. If we approach urban politics from this perspective, we are pulled firmly from a focus on municipal and state institutions either on the scalar or relational basis. In some degree this is a function of Rancière's attempt to reformulate democracy into a certain logic of social change that cannot be viewed as derived from institutional arrangements (i.e. parliamentary debate, free elections, majority voting, etc.). Simon Critchley (2007) outlines this relationship between state and politics as the following: 'If the activity of government continually risks pacification, order, security and what Rancière refers to as the "idyll of consensus", then politics consists in the manifestation of a dissensus that disturbs the order by which government wishes to depoliticize society' (129). For Critchley, politics therefore requires an anarchistic form and is, consequently, always at a distance from the formalities of state. Although Critchley takes this as a point of departure from Rancière in terms of viewing politics as multiple and everywhere, both view state policing as the reproduction of consensual logics that serves to mitigate dissensus.

Politics therefore emerges as a form of meta-theoretical imperative that diverts us from the visible workings of politics (i.e. state institutions, political parties). As such it dramatically repositions our perspective on the locale of politics. Specifically it draws our attention from the sites of policing (i.e. the state) and towards the multitude of equality claims. Critchley's (2007) 'democratic anarchism' certainly takes this position and withdraws from the state. Žižek presents the following criticism of such a withdrawal: '… if the space of democracy is defined by a distance toward the state, is Critchley not abandoning the field (of the state) all too easily to the enemy? That is to say: when Critchley defines today's constellation as one in which the state is here to stay, and in which we are caught in multiple displacements, and so on, the thesis is radically (and necessarily) ambiguous …' (2006: 333). Žižek's argument here is that if we look only to the radical margins for politics, we leave unchallenged the apparatus of the social. He continues by asking if Critchley's position does not unnecessarily evacuate the established means to generate radically different social relations:

> Does not Critchley's position, then, function as a kind of ideal supplement to the Third Way Left: a 'revolt' which poses no effective threat, since it endorses in advance the logic of hysterical provocation, bombarding the Power with 'impossible' demands, demands which are not meant to be met? Critchley is therefore logical in his assertion of the primacy of the Ethical over the Political: the ultimate motivating force of the type of political interventions he advocates is the experience of injustice, of the ethical unacceptability of the state of things. (Žižek 2006: 333–334)

Žižek clearly views the state as the place where hands get dirty; where difficult choices are made; where the left has to confront a possible utilization of state power. Without this difficult embrace of state-based action, Žižek is arguing that demands from the radical margins ultimately amount to (political) demands presented to class interests that will never be met (i.e. the request of the dissolving of class relations to the capitalist class).

This question of distance to the state/policing is clearly of concern for a re-politicized urban politics. Previous iterations of radical urban scholarship sought close relations to state power in order to attempt to generate different social relations. Current urban politics literatures continue to be concerned with the local state, but largely to understand its incapacities in the face of coercive competition. Rancière's post-democratic diagnosis is reflective of this literature, where he deems institutional politics to be a block on politics proper. Yet if we accept that politics only exists beyond the realms of the police, we are faced with the dilemma Žižek and Critchley have loudly debated. In particular, we must confront the question not only of where politics might emanate from, but also when these claims are evidenced what becomes of them once articulated and, specifically, what role the local/urban state has to play in this. The challenge can be thought of as two-fold, involving (i) a reconsideration of the utility of state power to radical politics and (ii) a confrontation with the associated problematic questions of power and hegemony.

The chapters in this section are all concerned with the constitution of community that comes before the state. That is to say, if we follow Rancière by understanding politics as emergent with unequal groups of people, then we first must be concerned with social groupings. With these identified, we hope the gap between community and state power that is evident in the urban politics literatures can be reduced. If our urban politics scholarship does not do this, then we risk re-inscribing those very exclusions that Rancière's theory of politics makes us attuned to. The community themes picked up on in the following chapters – sexuality, migrants, gender, class and environment – are far from exhaustive. Indeed we must leave for ever open our list of (potentially) marginalized social groups. Doing so is part of being attentive to the possibilities for dissensus.

In Chapter 8, Natalie Oswin describes the politics of sexuality that are a crucial part of Singapore's urban politics. Singapore has been held up around the globe as a hugely successful economic development story. In post-independence in 1965, Singapore has become a city with one of the highest per capita incomes in the world. Oswin describes how this story of economic success has contained specific notions of 'sexual citizenship'. Drawing on queer theory, Oswin explains how the city-state has constructed various sexual norms as part of its social agenda. These social agenda go beyond a simple promotion of heteronormativity and repression of homosexuality, incorporating a whole host of norms in, for example, family structure and social class. This complex social engineering all takes place on a backdrop of intensive economic strategizing and policy-making. What the latter connection illuminates is how community and urban regime agendas intertwine closely in Singapore.

In Jamie Winders' chapter our lens shifts towards immigrant communities in Nashville, Tennessee – Winders' concern is the growth of immigrant neighbourhoods in the city and their (dis-)connection with institutional politics. Here again we see the limits of an institutionally focused approach to urban politics. The problem posed in the chapter relates to how these immigrant communities actually come into being. Immigrants concentrate in particular neighbourhoods, but how is it that this propinquity relates to the constitution of community and then, presumably, political force? What the chapter illustrates is how the development of community and the related incorporation of immigrant communities into the political process is reliant on a broader shift in the imagination of the broader metropolitan community. In this case the re-imagination of metropolitan community, Nashville, is bound up with a discourse of multiculturalism.

We switch location again in Susan Hanson's chapter on female entrepreneurs. We go to Worcester, Massachusetts and Colorado Springs, Colorado to deconstruct the notion of the entrepreneur. In David Harvey's (1989) seminal commentary on the entrepreneurial city he describes 'entrepreneurialism' as a general ethos the shapes urban politics across much of the globe. For example he argues: 'Urban entrepreneurialism encourages the development of those kinds of activities and endeavours that have the strongest *localised* capacity to enhance property values, the tax base, the local circulation of revenues, and (most often as a hoped-for consequence of the preceding list) employment growth' (13). Whilst Harvey's (1989) reading of urban politics proved right, he does offer a sweeping notion of the entrepreneurial actions and actors. In Hanson's chapter we find by simply dividing urban entrepreneurs between genders, what being entrepreneurial means actually changes. Drawing on empirical data,

Hanson illustrates the varied ways in which female entrepreneurs constitute a particular constituency in urban politics.

The following chapter, written by Mark Davidson, focuses on the question of social class. While urban politics have long been understood as being dominated by politico-economic imperatives, in recent years we have been the declining relevance of class-based politics. Many of the working-class parties that once held local political power have now either disappeared or become much more centrist. Davidson explores this phenomenon and asks whether social classes are a relevant community in urban politics today. By using examples drawn from the UK, Davidson claims social class remains important to urban politics but that this community needs rethinking in an era of globalization. Specifically he calls for us to think about the external relations of the city as things that might need politicizing within the city.

Then in the final chapter of the section we switch gears to consider a thematically based community. In this chapter Matthew Huber examines the environmental component of urban politics, or rather he examines the urban component of environmental politics. Through a historical overview of the US environmental movement, Huber looks at the various ways in which the city itself has been represented. He claims the city has often been excluded from environmental politics, the city an anathema to nature. As a result, we find an urban politics lacking a concern for environment and an environmental politics lacking a concern for cities. The solution Huber proposes is to rethinking both the urban and nature, seeing them as mutually constituted. What this means for environmental movements and urban political communities is that their exclusion of each other's concerns cannot continue.

Collectively, then, the chapters in this section will take you across many different political terrains. To be sure they are not all connected. Rather the chapters all work to expose the varied constituencies of the urban political community. For example we might imagine that city populations in London and New York elect their mayors, but this sweeping description sometimes hides the varied communities that make up the city. Here are some questions to consider as you read on:

- Should certain communities have their claims elevated over those of others? And how do we mediate between different sets of claims emanating from different communities?
- What communities do you belong to? Do any of your community affiliations define your political interests and positions more than others?
- Do we all, at some basic level, belong to the same community?
- Do we always chose our community memberships or are they sometimes imposed on us?
- Does where you live impact your political interests? How much in common do you have with other people living in your city?

References

Critchley, Simon (2007) *Infinitely Demanding: Ethics of Commitment, Politics of Resistance*. Verso: London.
Harvey, D. (1989) 'From Managerialism to Entrepreneurialism: The Transformation in Governance in Late Capitalism', *Geografiska Annaler. Series B, Human Geography*, 71(1): 3–17.
Rancière, Jacques (1999) *Disagreement: Politics and Philosophy*. University of Minnesota Press: Minneapolis, MN.
Žižek, Slavoj (2006) *The Parallax View*. MIT Press: Cambridge, MA.

8

QUEERING THE CITY: SEXUAL CITIZENSHIP IN CREATIVE CITY SINGAPORE

Natalie Oswin

Introduction

Cross-disciplinary studies of homosexuality in western cities have been plentiful over the last few decades (Abraham, 2009; Chauncey, 1994; Houlbrook, 2005; Knopp, 1998; Podmore, 2001; Valentine, 1993) and various studies of gay and lesbian cultural politics in non-western cities have emerged more recently (Bacchetta, 2002; Martin, 2000; Oswin, 2005; Rofel, 1999; Tucker, 2009). It is thus by now well recognized that cities play a profoundly important role in facilitating the formation of gay and lesbian communities and movements around the world. It is also well recognized that gay and lesbian issues are important urban issues. Open any current text on critical urban politics (like this one) or urban social geography and you will find at least a mention of the relationship between homosexuality and the city. Or page through the recent volumes of urban studies journals and you will encounter a smattering of articles on this same theme (for example: Andersson, 2011; Bell and Binnie, 2004; Valentine and Skelton, 2003; Waitt and Gorman-Murray, 2011). We might say that there has been a 'queering' of urban studies insofar as the metropolitan lives, subcultures and social movements of gays and lesbians are now seen as valid objects of study.[1] But beyond its scholarly and popular use as a synonym for sexual minorities, 'queer' signifies a second meaning. It refers to a poststructural theoretical approach that challenges the presumed fixity of sexual identities and critiques the ways in which sexual norms are deployed as part of broad structures of governance. In this chapter, 'queer' is used in both of these senses. Gays and lesbians, or 'queers', in the city are discussed to some extent. But the main emphasis is on setting out what queer theory can bring to critical understandings of urban politics beyond an awareness of the inequities of the heterosexual–homosexual binary. The emphasis, in other words, is on taking the 'queering' of urban studies further, to expand understanding of the range of subjects who are rendered abject, abnormal, or 'queer' in urban heteronormative sites.

In what follows, I first set out the difference between 'gay and lesbian studies' approaches that emerged in the 1970s and 1980s to study minority sexual identities and the 'queer' theoretical critique that emerged in the 1990s. The latter problematizes the notion of essential or fixed sexual identities and forces us to go beyond an understanding of urban spaces as either 'straight' or 'gay' in order to consider the ways in which sexual norms are constructed in place. While queer theory has a broad reach across the social sciences and humanities, I focus on its uptake within the discipline of geography since geographers have been at the forefront of work on the urban politics of sexuality. Then, to ground the abstractions of a queer theoretical approach and convey its utility for the study of urban politics, I turn to a case study from Singapore. Within this city-state, there has been an unprecedented liberalization of the government's approach to the regulation of homosexuality in recent years. This change is part of the city-state government's efforts to foster a new, post-industrial, creative economy in order to stay ahead in the current competitive global economic environment. Considering the limits of this liberalization, I critique Singapore's creative/global city project as still fundamentally heteronormative. Further, I argue that this critique is by no means of relevance only to a small constituency of sexual minorities by examining the ways in which Singapore's heteronormativity is tied in fundamental ways to broad forms of social polarization in this newly entrepreneurial city. Understanding public debates over homosexuality's place in Singapore as only one part of a broader story of sexual citizenship, I consider the central role that a particular ideal of the proper family has played in the city-state government's efforts to foster modernity and progress in the independence era. Through an examination of the differential policies directed at the two main categories of migrant labourers – 'foreign workers' and 'foreign talent' – upon which Singapore's creative city strategy relies, I consider the ways in which not only certain expressions of homosexuality but also certain expressions of heterosexuality are marginalized in this global city-state. Thus I argue that we need a broader queer approach if we are to fully understand, and critically respond to, the exclusions of sexual politics in the city.

Theorizing sexual subjects in the city

Within the vast interdisciplinary field of sexuality studies, geographers have been offering unique insights into the relationship between sexuality and space since the late 1970s. Not surprisingly, much of this scholarship examines the particular relationship between sexuality and urban space. In this section, I briefly trace the emergence of studies of 'gay and lesbian' and then 'queer' geographies, highlighting the specific epistemological underpinnings of each and thus their quite different approaches to urban sexual politics. While both frameworks offer important insights and avenues for research, I emphasize the ways in which the 'queer' critique of the notion of sexual identity opens the political project of urban sexuality studies to the necessity of interrogating a much wider range of sexual subjects in the city than a narrower 'gay and lesbian' approach allows.

Geographers began to explore expressions of sexualities in urban areas with scattered studies of gay social spaces in US cities emerging in the late 1970s and early 1980s (Weightman, 1980). Though this very early work was largely descriptive and empirically focused, scholarship informed by feminist theory and political economy appeared in the mid-1980s to address urban gay politics and the role of gay men in processes of gentrification (Lauria and Knopp, 1985; Knopp, 1992). Work on lesbian geographies ensued later (and less abundantly) and laid the foundation for understandings of lesbian social networks and residential experiences (Peake, 1993; Rothenberg, 1995). These early explorations of gay and lesbian spaces in cities greatly extended our knowledge of urban social geography by calling attention to the lives of non-heterosexuals as valid subjects of geographical enquiry and putting the study of sexuality on the disciplinary agenda.[2] But the emergence of queer theory within literary theory in the early 1990s and its uptake within geography (and other social sciences) soon thereafter led to the articulation of some foundational critiques of these gay and lesbian urban geographies which significantly re-oriented sexuality and space studies. I now turn to a brief discussion of the key tenets of queer theory before elaborating the ways in which these queer insights have led to the advancement of new geographical understandings of urban sexual politics.

As a theoretical approach, 'queer' offers a potent critique of identity politics and thus of much gay and lesbian studies work of the 1970s and 1980s. Whereas a gay and lesbian studies approach tends to understand sexual identities to be natural, fixed and biologically determined, queer theorists assert that sexual identities are social constructions that do not pre-exist their worldly (i.e. cultural and linguistic) deployments.[3] Attesting that sexualities are performed, that they are something that we *do* rather than something that we *have*, queer theory understands sexual identity as unfixed and constantly changing. These anti-essentialist understandings of sexual identities owe much to the post-structuralist theorist Judith Butler. In *Gender Trouble* (1990) and *Bodies that Matter* (1993), Butler offers a profound critique of the presumption that sex and sexuality are biologically driven or inherent traits. She argues that we are not born as 'women' and 'men' or 'girls' and 'boys' but rather are interpellated as gendered beings from the day we are born. 'Gendering' as an active process entails the taking on of gender norms through a performative doing. As Butler explains, performativity operates through 'the reiterative power of discourse to produce the phenomena that it regulates and constrains' (1993: 2). So the subject is brought into being through discourse and the gendered self has no ontological status apart from the acts of discourse that compose it. Further, just as 'sex' is a regulatory fiction, something that is learned and performed rather than something 'natural', so too is sexuality; with the norm of compulsory heterosexuality playing a key part in regulating gender as a binary relation.

Suggesting that one never 'is' heterosexual or homosexual, but is only ever in a condition of 'doing' or 'performing' heterosexuality or homosexuality, Butler thus opens up political possibilities by re-orienting our understanding of sexuality from 'thing' to 'process' and acknowledging that sexual identity categories are malleable. But, crucially, she insists that they are not malleable in the abstract since the performance of identity is not voluntarist. Sexual identities cannot simply be changed at will, Butler argues, since although sexuality 'is not a noun … neither is it a set of free-floating attributes' (1990: 33). As such, Butler thus effectively respatializes our understanding of subjectivity by suggesting that the subject comes into being in the world rather than outside and against it. While she argues that discourse is not fixed, she also notes that social constraints place limits on the expression of sexual norms. In other words, sexual norms and identities are not easily changeable because they are constrained by context. So the critical task at hand is to challenge the myriad processes that make the contingent phenomenon of hegemonic heterosexuality seem fixed.

On this point, the queer reorientation of sexuality and space studies within geography comes into view. For the contingencies of hegemonic heterosexuality come to appear natural only in place. The 'normal' and the 'perverse' are not biologically determined things, but are ideas that are geographically and historically specific. Taking on board these anti-essentialist understandings of sexual identity, geographers since the mid-1990s have done much more than locate gay and lesbian spaces by turning their attention to the ways in which sexual identities and spaces are performatively and discursively produced. As queer theory challenges sexual categories as natural, so has queer geographical scholarship come to focus on how sexualized spaces are in fact socially constructed and geographically contingent. Rather than simply mapping gay and/or lesbian spaces in the city, geographers now also ask how certain spaces come to be identified as 'gay' or 'lesbian' and question whether these categories are in fact discrete. Further, rather than accepting 'straight' spaces as natural or normal, they call attention to heterosexuality and its spatial manifestations as actively produced and ideologically maintained. In short, by taking up the queer insight that sexual identities are socially constructed power relationships rather than essential biological traits, queer geographers recognize that all spaces in cities are sexualized, in an active sense, and that sexual subjectivities are performed in context-specific ways.

So a queer geographical approach to urban sexual politics seeks to render 'normal' sexuality strange rather than to identify and delineate 'straight' and 'queer' urban spaces. It expands our understandings of where sexual politics is found in the city and through which bodies it is enacted. As such, a 'queered' urban studies in its fullest iteration interrogates not just the identity politics of homosexuality; it interrogates heteronormativity, the cultural logic that makes particular expressions of heterosexuality seem right. To convey the utility of these abstract claims more concretely, I now turn to a case study that puts them to work. In the following analysis of sexual citizenship in contemporary Singapore, I build on the insight that space is actively sexualized and examine the ways in which a heteronormative logic does much more than police a heterosexual–homosexual binary.

The heteronormative politics of 'talent' in global city Singapore

The Southeast Asian island of Singapore emerged from a long period of British colonial rule to become an independent city-state in 1965. Since then, its socioeconomic development has been nothing short of extraordinary and Singapore has gained recognition as a leading 'global city'. Maintaining this status in a changing global environment, however, has required a drastic economic reorientation in recent years. To deal with declining manufacturing activity, state-led development plans have fostered a shift to a post-industrial, knowledge-based economy since the mid-1990s. This strategy can certainly be considered a success if measured solely in terms of economic growth. But many scholars concerned with the social and political aspects of Singapore's new economy offer quite critical appraisals (Chang, 2000; Goh, 2005; Ho, 2006; Sparke et al., 2004; Wong and Bunnell, 2006). In short, and in resonance with common critiques of other 'global' and 'creative' city projects around the globe, Singapore's developmental efforts have been characterized as top-down, hierarchical, narrowly economistic and, most importantly, socially polarizing. These critiques of Singapore's urban politics are immensely valuable. But I'll argue in this section that they are incomplete. For while this existing literature attends to schisms along the lines of race, class, gender and nationality in contemporary Singapore, it does not consider inequities based on sexuality. Most obviously, the exclusion of homosexuals within Singapore's polity is left out of existing critical discussion. But as I will demonstrate, the exclusionary effects of heteronormativity in the city-state extend to many more than gays and lesbians/'queers' and thus a queer reading can deepen understanding of the causes of and necessary political responses to social polarization in Singapore.

In 2003, an article based on an interview with Singapore's then Prime Minister Goh Chok Tong appeared in *Time Asia* magazine. Entitled 'The Lion in Winter', the article covered the Singapore government's efforts to shake off the city-state's authoritarian image in order to foster an entrepreneurial, creative spirit in the face of changing global economic conditions. It includes the following example of resulting shifts in Singapore society:

> The city now boasts seven saunas catering almost exclusively to gay clients ... something unthinkable even a few years ago ... Prime Minister Goh says his government now allows gay employees into its ranks, even in sensitive positions. The change in policy, inspired at least in part by the desire not to exclude talented foreigners who are gay, is being implemented without fanfare, Goh says. 'We are born this way and they are born that way, but they are like you and me.' (Elegant, 2003)

This was an out of the blue and unprecedented government statement and, in short, the government's approach to public expressions of homosexuality has since been liberalized as the Singapore government has basically adopted Richard Florida's (2002) mantra that 'tolerance' attracts 'talent' (see Box 8.1). Since Goh's 2003 interview (Elegant, 2003), queer cultural production has flourished in Singapore: local gay film and theatre productions have been able to emerge because of loosened censorship restrictions; police raids that were not uncommon in the 1990s have ceased and bars, saunas, and other gay and lesbian owned business have been allowed to operate without interference; and a large gathering of gays and lesbians that organizers have named 'Pink Dot' has been allowed to take place in a public park annually since 2009 (see Figure 8.1).[4] So much has changed that Singapore has come to be known popularly as Asia's 'new gay capital'.

Figure 8.1 Pink Dot formation, Singapore (source: Pink Dog SG with permission)

BOX 8.1 RICHARD FLORIDA'S CREATIVE CLASS THESIS

Economic geographer and urban policy advisor Richard Florida argues that 'creative people' – his term for the higher order service sector workforce that drives new, 'creative' urban economies – power urban and regional economic growth and that they 'prefer places that are diverse, tolerant and open to new ideas' (2002: 249). Florida argues that cities must not only create the right market/ economic conditions for growth. They must also create the right non-market/ quality of life conditions by fostering his '3T's' of economic development – technology, talent and tolerance. He states, 'creative people are attracted to, and high-tech industry takes root in, places that score high on our basic indicators of diversity – the Gay, Bohemian and other indexes ... Why would this be so? It is not because high-tech industries are populated by great numbers of bohemians and gay people. Rather,

(Continued)

> *(Continued)*
>
> artists, musicians, gay people and the members of the Creative Class in general prefer places that are open and diverse' (2002: 250).
>
> Various urban studies scholars have challenged Florida's argument on several fronts. These include methodological and empirical critiques, political/ social equity concerns and questions about the reproducibility of 'creative city' strategies (see Luckman et al., 2009; Markusen, 2006; Peck, 2005). More directly related to my immediate empirical concerns, his 'gay index' has come under specific scrutiny. As David Bell and Jon Binnie (2004) note, Florida relies on census data on gays and lesbians in same sex partnerships to make his linkage between the presence of 'gays' and 'creativity' and 'the gay index is therefore an index of respectability, of nicely gentrified neighbourhoods' (p. 1817). Further, they argue that the incorporation of 'sexual others' into entrepreneurial urban governance strategies of place promotion 'has meant tightening regulation of the types of sexualized spaces in cities' (p. 1818).

Though these are remarkable changes, the government has also made very clear that there are absolute limits to this liberalization. Permits have not been granted for various public talks relating to gay and lesbian rights, the gay and lesbian activist organization People Like Us has been refused registration as a society, and perhaps most significantly a 2007 lobbying effort to repeal Section 377A of the Penal Code, a statute that criminalizes 'gross indecency' between men, was dismissed. As Eng-Beng Lim has argued, the Singapore state has a 'new – if volatile – attitude toward its queer citizenry: we'll leave you alone so long as homosexuality is not encouraged and is no more than a marketplace commodity' (2005: 297–298).[5] In other words, 'tolerance' for the sake of attracting 'talent' has not fundamentally changed the legal position of gays and lesbians in the city-state. Some gays and lesbians who are seen to be contributing to the new economy are allowed to tentatively step into the public sphere (for now at least, as long as value is tied to 'creativity' and the creation of urban 'buzz'). But enfranchisement is off the table.

The Singapore government's current position on homosexuality was most directly expressed in Prime Minister Lee Hsien Loong's (2007) speech to parliament that explained why the sodomy law has been maintained. In this speech, Lee declared that the gay and lesbian community, 'include[s] people who are responsible and valuable, highly respected contributing members of society' and 'among them are some of our friends, our relatives, our colleagues, our brothers and sisters, or some of our children'. He prescribed that 'they too must have a place in this society, and they too are entitled to their private lives' while nonetheless insisting that, 'homosexuals should not set the tone for Singapore society'. Lee further laid bare a stark terrain with the declaration that, 'there is space, and there are limits'. While Singaporeans were encouraged by their head of state to adopt a 'live and let live attitude' in regard to gays and lesbians, he asserted that 'the overall society ... remains conventional, it remains straight' because:

> the family is the basic building block of this society. It has been so and by policy, we have reinforced this, and we want to keep it so. And by family in Singapore we mean one man, one woman marrying, having children and bringing up children within that framework of a stable family unit ... It is not so in other countries, particularly in the West anymore, but it is here.

The exclusions of Singapore's global/creative city project extend quite obviously to gays and lesbians and critical scholarship needs to incorporate a queer lens on this front alone. But there is even more at stake, as we can see if we critically interrogate the linkage that Lee ultimately makes between Singapore's 'straightness' and the family. For it is a very specific heterosexual nuclear family upon which the state relies to facilitate Singapore's urban and national development. On this point, my argument that heteronormativity is central to Singapore's creative/global city project in much more widespread ways than a sole focus on gay and lesbian disenfranchisement allows us to understand comes into view. I want to now shift to this broader point to explain that not just gays and lesbians, but many people that fit into the category 'heterosexual' are in fact rendered out of place in Singapore. In other words, I will point to some of the ways in which not just gays and lesbians are 'queered' by the specific notion of the proper family to which PM Lee's 2007 speech only gestures.

Throughout Singapore's independence period, economic planning has gone hand in hand with comprehensive social engineering efforts. Notably, attempts to shape the institution of the family have been at the core of these extensive governmental interventions into social life. Building on social policies meant to spur on Singapore's development in the late colonial period, as soon as the People's Action Party government (which is still in power today) took office, it began to put in place extraordinary measures to 'upgrade' the family as part of its drive to modernize the city-state.[6] These efforts included aggressive family-planning campaigns and anti-natalist measures guided by overtly eugenicist ideas. But these anti-natalist measures worked all too well and have been replaced since the early 1980s by pro-natalist initiatives aimed at particular Singaporeans (Drakakis-Smith et al., 1993). Along with constant exhortations to marry emanating from the state-controlled media, various government ministries and a battery of state-supported organizations, a range of policy initiatives that implore educated, professional people to have as many children as possible while encouraging those without considerable means and/or educational qualifications to continue to limit their family sizes have been put in place over the last thirty years. These efforts have made it abundantly clear that the Singapore state envisions a very particular sort of family as central to accomplishing its developmental aims.

Despite these truly colossal state efforts to build and grow the city-state's families in these particular ways, the intended results have not been achieved and Singapore's fertility rate has been below replacement level for many years. This latter fact is a source of much official anxiety since Singapore's government has stated that the achievement of its long-term economic growth targets requires a considerable population increase. As such, while continuing to promote marriage and procreation, in 2006 a new strategy for increasing the population was advanced. PM Lee (2006) signalled this strategy overtly in his National Day Rally speech as follows:

> If we want our economy to grow, if we want to be strong internationally, then we need a growing population and not just numbers but also talents in every field in Singapore ... there are things which we can do as a government in order to open our doors and bring immigrants in. But more importantly as a society, we as Singaporeans, each one of us, we have to welcome immigrants, welcome new immigrants.

Of course, postcolonial Singapore has long relied on migration flows to meet its human resource demands. But what has been unique since 2006 is the government's interest in 'welcoming new immigrants' by encouraging immigrants to become Singapore citizens.[7]

Recent policies concertedly encourage the naturalization of *suitable* immigrants and aim to welcome 'talent' into the national family en masse. For instance, the Immigration and Checkpoints Authority produced a brochure titled 'Embrace Singapore, Where you Belong!' (2007), which declares: 'Share your talents. Build a nation.' In a section of this brochure addressed 'Dear Permanent Residents', the fact that the government now considers migrants an essential part of Singapore's social reproduction is expressed as follows:

> Many of you have even brought family members to Singapore to help you better adjust to living here ... You and your family members have benefited from what Singapore has to offer, just as Singapore has progressed and prospered with your labour and contributions. It is time to take a step further and become a part of the Singapore family. (See also Economic Strategies Committee, 2010)

Another section of the brochure overtly targets 'family members' by detailing what Singapore has to offer to 'foreigners and their families'. As a result of this shift in policy, the number of new Permanent Residents and citizens in Singapore has swelled in the last couple of years. But what I wish to highlight is that this policy towards 'foreign talent' contrasts strikingly with the policy framework that manages temporary 'foreign workers' – the other, and much larger, group of migrants that Singapore relies upon. In 2009, Singapore's foreign population numbered 1.2 million, approximately a quarter of its total population. Within this group only around 170 000 are expatriate professionals (i.e. 'foreign talent') hailing from a range of countries including China, India, the United Kingdom and Australia, while the remainder are low-wage foreign workers from countries such as the Philippines, Indonesia, Thailand, India, Bangladesh and Sri Lanka (Baey, 2010). There are no glossy brochures inviting this latter group into the national family. Rather, they are rendered permanently transient and alien through policies that regulate their time in Singapore.[8]

Many scholars have offered very useful critiques of the distinctions made between 'foreign talent' and 'foreign workers' in Singapore and of the different policy frameworks that govern the lives of each group (Poon, 2009; Teo and Piper, 2009; Yeoh and Chang 2001).[9] Existing scholarship focuses on gender, class and race dynamics to explain this bifurcated labour regime. I argue that the sexual regulation of these two groups of migrants is also key to their differentiation. For while those migrants who fall under the category 'foreign talent' are invited into the national family, those who are characterized as 'foreign workers' are excluded from Singapore society and rendered alien to the city-state. The Employment of Foreign Workers Act states:

> the foreign worker shall not go through any form of marriage or apply to marry under any law, religion, custom or usage with a Singapore Citizen or Permanent Resident in or outside Singapore ... If the foreign worker is a female foreign worker, the foreign worker shall not become pregnant or deliver any child in Singapore during the validity of her Work Permit/Visit Pass. The foreign worker shall not indulge or be involved in any illegal, immoral or undesirable activities, including breaking up families in Singapore.

While 'foreign talent' is being welcomed into the city-state because they are seen to be 'quality' additions to the population, 'foreign workers' are cast as external to the proper Singaporean family and are subjected to a series of regulatory measures that aim to ensure that they remain single for the duration of their stay in Singapore. In other words, while 'foreign talent' fit into the heteronormative ideal that has long underpinned Singapore's economic and social development, 'foreign workers' must exist outside the familial norm or face deportation.[10]

Through this example we can see that the 'straightness' of Singapore's urban and national space is selective and that heterosexual privilege is not extended to all heterosexuals. When we consider the range of ways in which family, kinship and domesticity have been central areas of governmental intervention throughout Singapore's postcolonial era, it becomes apparent that not just gays and lesbians have been 'queered' in the city-state. To fulfil a particular vision of the 'creative' and 'global' city, population policies regulate a heteronormative ideal that keeps all those who do not fit into the notions of 'quality' citizens and proper family on the margins. While I have elaborated this argument most specifically in relation to the sexual regulation of 'foreign workers', it has even further reach. There are of course still other heterosexual 'others' who are rendered abnormal and abject by heteronormativity in Singapore. This broader groups of sexual 'others' include, for instance, single parents who are not entitled to housing subsidies, single mothers who are not entitled to full maternity benefits, persons without university education who are discouraged from having more than two children through a series of financial disincentives, and sex workers who face much social opprobrium although their work is technically legal. Sexual politics in this city, therefore, affects a much wider group of subjects than a focus on the oppositional politics of homosexuality versus heterosexuality allows.

Conclusion

In this chapter, I have sought to demonstrate that a queer approach offers a critique of the ways in which sexual identity categories are deployed in the city. Whereas early 'gay and lesbian' approaches – particularly as taken up within geography – tended to rely on essentialist understandings of sexual identities and to focus on describing expressions of these identities in discrete spaces in the city, a queer approach suggests that sexual identities are socially constructed and must be interrogated for their historical and geographical contingencies. By questioning the naturalness of sexual identities, queer geographical scholarship broadly and deeply challenges the heteronormality of urban spaces. It opens up understandings of where sexual politics occur in the city and through which bodies. As the Singapore case that I have outlined above demonstrates, critical approaches to urban politics ought to account for the exclusionary politics of sexual citizenship. Incorporating queer theoretical insights helps us to do so in two ways. First, queer approaches put the exclusions of sexual minorities such as gays and lesbians on the agenda. Second, they call attention to the ways that sexual norms do more than pit a monolithic hegemonic heterosexuality against a marginalized homosexuality. Within urban space, there are multiple heterosexualities with some expressions of these identities being valued more highly than others. As such, a narrow emphasis on heterosexual versus homosexual politics in the city misses much. Queer theory reveals the nuanced workings of heteronormativity as a sexualized process that cannot be extricated from gendered, racialized and classed (among others) processes in the city. Since sexual politics are fractured, contingent and have broad effects, there is much need for queer critical response. Alongside and in relation to analyses of patriarchy, racism and capitalism, critical approaches to urban politics must examine heteronormativity in order to fully grasp the ways in which certain lives are rendered precarious in the city.

Notes

1. Trans issues have unfortunately not been taken up substantively. But see Doan (2011).
2. This early scholarship on gay and lesbian geographies faced significant resistance from some established geographers and was truly path-breaking (see Brown and Knopp, 2003).
3. These neat opposing typologies of 'gay and lesbian' and 'queer' approaches are admittedly caricatures to a certain extent. Some scholars associated with gay and lesbian studies of the 1970s and 1980s such as John D'Emilio and Jeffrey Weeks were in fact among the first to advance social constructionist approaches to the study of sexuality. See Jagose (1996) for detailed discussion of these two approaches and their interrelations.
4. See Russell Heng (2001) for an account of the historical treatment of homosexuality in Singapore and the emergence of its 'gay community'.
5. See also Yue (2007) for an analysis of the links between the Singapore government's creative city aspirations and the flourishing of queer cultural production.
6. See Salaff (1988) for a detailed account of the early postcolonial government's efforts to 'upgrade' the family and see Oswin (2010a, 2010b) for analyses of the links between Singapore's colonial and postcolonial family policies.
7. As a city-state, the urban politics of immigration found in Singapore are fairly unique. But urban scholars in cities around the globe have been increasingly calling attention to immigration and citizenship as urban political questions. See, for instance, Coleman (2012), Dikeç (2007) and Winders (2008).
8. Though word limits preclude a full examination of constraints that migrant workers experience while in Singapore, it must be noted that these include significant spatial constraints. Domestic workers must live in their employers' homes and no time off is legally mandated. Those working in the construction sector are required to live in dormitories that are set apart from Singaporean residential spaces and they travel to and from work in specially chartered lorries.
9. Beyond Singapore studies, there is a considerable body of scholarly work that addresses the lives of migrant workers in various cities around the globe. Much of this work advances a feminist critique, examining the particular challenges faced by female migrant workers due to gender biases and challenging the devalorization of migrant labour as 'feminized' labour. These are important insights, but I argue that in addition to challenging the patriarchal underpinnings of inequitable migrant labour regimes, we must also interrogate their

heteronormative biases. For migrant labourers, whether female or male, in numerous cities around the globe are excluded from urban citizenship by policies that render them impermanent and outside the national family.

10 Two qualifications are in order here. First, 'foreign workers' in Singapore of course exert agency and many manage to have sexual lives while in the city-state. Second, I am not suggesting that all migrants are heterosexual. Rather, my point is that the Singapore state presumes them to be and regulates them as such.

References

Abraham, Julie (2009) *Metropolitan Lovers: The homosexuality of cities*. University of Minnesota Press: Minneapolis, MN.

Andersson, J. (2011) 'Vauxhall's post-industrial pleasure gardens: "Death wish" and hedonism in 21st-century London', *Urban Studies*, 48(1): 85–100.

Bacchetta, P. (2002) 'Rescaling transnational 'queerdom': Lesbian and "lesbian" identitary-positionalities in Delhi in the 1980s', *Antipode*, 34(5): 937–973.

Baey, Grace (2010) 'Borders and the exclusion of migrant bodies in Singapore's global city-state', MA thesis, Queens University, Department of Geography.

Bell, D. and Binnie, J. (2004) 'Authenticating queer space: Citizenship, urbanism and governance', *Urban Studies*, 41(9): 1807–1820.

Brown, M. and Knopp, L. (2003) 'Queer cultural geographies – we're here! We're queer! We're over there, too!', in Anderson, K., Domosh, M., Thrift, N. and Pile, S. (eds) *Handbook of Cultural Geography*. Sage, London: pp. 313–324.

Butler, Judith (1990) *Gender Trouble: Feminism and the subversion of identity*. Routledge: New York.

Butler, Judith (1993) *Bodies That Matter: On the discursive limits of 'sex'*. Routledge: New York.

Chang, T.C. (2000) 'Renaissance revisited: Singapore as a global city for the arts', *International Journal of Urban and Regional Research*, 24(4): 818–831.

Chauncey, George (1994) *Gay New York: Gender, urban culture and the making of the gay male world 1890–1940*. Basic Books: New York.

Coleman, M. (2012) 'The local migration state: The site-specific devolution of immigration enforcement in the U.S. South', *Law & Policy*, 34(2): 159–190.

Dikeç, Mustafa (2007) *Badlands of the Republic: Space, politics and urban policy*. Wiley-Blackwell: Oxford.

Doan, P. (2011) 'Queerying identity: Planning and the tyranny of gender', in Doan, P. (ed.) *Queerying Planning: Challenging heteronormative assumptions and reframing planning practice*. Ashgate: Burlington, VT. pp. 89–106.

Drakakis-Smith, D., Graham, E., Teo, P. and Ooi, G.L. (1993) 'Singapore: Reversing the demographic transition to meet labour needs', *Scottish Geographical Magazine*, 109(3): 152–163.

Economic Strategies Committee (2010) *Report of the Subcommittee on making Singapore a leading global city*. Accessed at: http://app.mof.gov.sg/data/cmsresource/ESC%20Report/Subcommittee%20on%20Making%20Singapore%20a%20Leading%20Global%20City.pdf.

Elegant, S. (2003) 'The lion in winter', *Time Asia*, 30 June. Accessed 20 October 2010 at: www.time.com/time/asia/covers/501030707/sea_singapore.html.

Florida, Richard. (2002) *The Rise of the Creative Class: And how it's transforming work, leisure, community and everyday life*. Basic Books: New York.

Goh, Robbie (2005) *Contours of Culture: Space and social difference in Singapore*. Hong Kong University Press: Hong Kong.

Heng, R. (2001) 'Tiptoe out of the closet: The before and after of the increasingly visible gay community in Singapore', *Journal of Homosexuality*, 40: 81–97.

Ho, E. (2006) 'Negotiating perceptions of belonging and citizenship in a transnational world: Singapore, a cosmopolis', *Social and Cultural Geography*, 7(3): 385–401.

Houlbrook, Matt (2005) *Queer London: Perils and pleasures in the sexual metropolis, 1918–1957*. University of Chicago Press: Chicago, IL.

Immigration and Checkpoints Authority (2007) 'Embrace Singapore, where you belong!' Singapore.

Jagose, Anna Marie (1996) *Queer Theory: An introduction*. New York University Press: New York.

Knopp L. (1992) 'Sexuality and the spatial dynamics of capitalism', *Environment and Planning D: Society and Space*, 10(6): 651–669.

Knopp L. (1998) 'Sexuality and urban space: Gay male identity politics in the United States, the United Kingdom and Australia', in Fincher, R. and Jacobs, J.M. (eds) *Cities of Difference*. Guilford Press: New York. pp. 149–176.

Lauria, M. and Knopp, L. (1985) 'Toward an analysis of the role of gay communities in the urban renaissance', *Urban Geography*, 6(2): 152–169.

Lee H.L. (2006) National Day Rally Speech. Singapore Government Press Release. Ministry of Information, Communications and the Arts: Singapore.

Lee H.L. (2007) Speech to Parliament on reading of Penal Code (Amendment) Bill, 22 October. Singapore Government Press Release. Ministry of Information, Communications and the Arts: Singapore.

Lim, E.B. (2005) 'The Mardi Gras boys of Singapore's English-language theatre', *Asian Theatre Journal*, 22(2): 293–309.

Luckman, S., Gibson, C. and Lea, T. (2009) 'Mosquitoes in the mix: How transferable is creative city thinking?', *Singapore Journal of Tropical Geography*, 30(1): 70–85.

Markusen, A. (2006) 'Urban development and the politics of a creative class: Evidence from a study of artists', *Environment and Planning A*, 38(10): 1921–1940.

Martin, F. (2000) 'From citizenship to queer counterpublic: Reading Taipei's new park', *Communal/Plural*, 8(1): 81–94.

Oswin, N. (2005) 'Researching "gay Cape Town", finding value-added queerness', *Social and Cultural Geography*, 6(4): 567–586.

Oswin, N. (2010a) 'The modern model family at home in Singapore: A queer geography', *Transactions of the Institute of British Geographers*, 35(2): 256–268.

Oswin, N. (2010b) 'Sexual tensions in modernizing Singapore: The postcolonial and the intimate', *Environment and Planning D: Society and Space*, 28(1): 128–141.

Peake, L. (1993). '"Race" and sexuality: challenging the patriarchal structuring of urban social space?', *Environment and Planning: Society and Space D*, 11(4): 415–432.

Peck, J. (2005) 'Struggling with the creative class', *International Journal of Urban and Regional Research*, 29(4): 740–790.

Podmore, J. (2001) 'Lesbians in the crowd: gender, sexuality and visibility along Montreal's Boulevard St-Laurent', *Gender, Place and Culture*, 8(4): 333–355.

Poon, A. (2009) 'Pick and mix for a global city: Race and cosmopolitanism in Singapore', in Goh, D., Gabrielpillai, M., Holden, P. and Khoo, G.C. (eds) *Race and Multiculturalism in Malaysia and Singapore*. Routledge: New York. pp. 70–85.

Rofel, L. (1999) 'Qualities of desire: Imagining gay identities in China', *GLQ*, 5(4): 451–474.

Rothenberg, T. (1995) 'And she told two friends': Lesbians creating urban social space', in Bell, D. and Valentine, G. (eds) *Mapping Desire: Geographies of Sexualities*. Routledge: New York. pp. 165–181.

Salaff, J.W. (1988) *State and Family in Singapore: Restructuring a developing society*. Cornell University Press: Ithaca, NY.

Sparke, M., Sidaway, J., Bunnell, T. and Grundy-Warr, C. (2004) 'Triangulating the borderless world: Globalisation, regionalisation and the geographies of power in the Indonesia–Malaysia–Singapore growth triangle', *Transactions of the Institute of British Geographers*, 29(4): 485–489.

Teo, Y.Y. and Piper, N. (2009) 'Foreigners in our homes: Linking migration and family policies in Singapore', *Population, Space and Place*, 15(2): 147–159.

Tucker, Andrew (2009) *Queer visibilities: Space, identity and interaction in Cape Town*. Blackwell: Oxford.

Valentine, G. (1993) '(Hetero)sexing space: Lesbian perceptions and experiences of everyday spaces', *Environment and Planning D: Society and Space*, 11(4): 395–413.

Valentine, G. and Skelton, T. (2003) 'Finding oneself, losing oneself: The lesbian and gay "scene" as a paradoxical space', *International Journal of Urban and Regional Research*, 27(4): 849–866.

Waitt, G. and Gorman-Murray, A. (2011) 'Journeys and returns: Home, life narratives and remapping sexuality in a regional city', *International Journal of Urban and Regional Research*, 35(6): 1239–1255.

Weightman, B. (1980) 'Gay bars as private places', *Landscape*, 24: 9–17.

Winders, J. (2008) 'An "incomplete" picture? Race, Latino migration, and urban politics in Nashville, Tennessee', *Urban Geography*, 29(3): 246–263.

Wong, K.W. and Bunnell, T. (2006) '"New economy" discourse and spaces in Singapore: a case study of One North', *Environment and Planning A*, 38(1): 69–83.

Yeoh, B.S.A. and Chang, T.C. (2001) 'Globalising Singapore: Debating transnational flows in the city', *Urban Studies*, 38(7): 1025–1044.

Yue, A. (2007) 'Creative queer Singapore: The illiberal pragmatics of cultural production', *Gay and Lesbian Issues and Psychology Review*, 3(3): 149–160.

9

MAKING SPACE IN THE MULTICULTURAL CITY: IMMIGRANT SETTLEMENT, NEIGHBOURHOODS AND URBAN POLITICS

Jamie Winders[1]

Introduction

> The everyday life-spaces of the city – its neighborhoods, parks, streets, and buildings – are ... both the medium through which citizenship struggles take place and, frequently, that which is at stake in the struggle. (Secor, 2004: 353)

As 'a defining social context of American life' (Campbell et al., 2009: 461; see also Park et al., 1925; Yarbrough, 2003), neighbourhoods are key to how urban geographies and politics function at a number of scales (Ley, 1995; Amin, 2002; Martin, 2003b). Through neighbourhoods, residents situate their lives vis-à-vis broader urban practices, from economic (dis)investment to political representation (Rocco, 1997; Martin, 2003b), using their experiences in the neighbourhood to lobby for changes at wider scales (Jackson, 2008; Heynen, 2009). For urban scholars, the neighbourhood, and especially its changing racial and ethnic composition, has been a way, and a space through which, to examine immigrant assimilation (Brubaker, 2001; Wright et al., 2005), residential segregation (Massey and Denton, 1993; Peach, 1996; Ellis and Wright, 1998), racialization (Anderson, 1987), and other themes. In all these ways, for urban residents and scholars alike, assessing the city often means examining its neighbourhoods.

Despite neighbourhood's import, it remains a messy construct (Martin, 2003a). The composition of neighbourhoods change over time, as do their boundaries and public images. Moreover, the idea of 'neighbourhood' itself means different things to different people. Deborah Martin (2003a) has argued that 'neighbourhood' often comes into being as a space and a construct 'through political strategy and contestation' over its spaces (362), as residents organize around and through the neighbourhood as an idea(l). This chapter builds on that argument, by suggesting that 'neighbourhood' is also produced through struggles over its place within wider urban politics in moments of transition. As this chapter will show, change itself can be key to the production, *and contestation*, of neighbourhood as a space of social identity and political representation and can reconfigure the connection between the space of the neighbourhood and the sphere of urban politics. When neighbourhoods are in transition, they place in sharp relief the scalar linkages between changes across the street and wider circuits of political visibility in the city.

To develop this argument about the relationship between neighbourhood change and urban politics, this chapter examines the saga of changing neighbourhoods in Nashville, Tennessee. Since the mid-1990s, many southern US cities like Nashville have experienced immigrant, especially Latino, settlement for the first time in recallable history (Winders, 2005; Massey, 2008). As Latino *workers* initially drawn to southern cities' labour markets have settled in their neighbourhoods as *residents*, urban politics in southern cities have been transformed through new racial dynamics (McClain et al., 2006; Winders, 2008), new cultural tensions (Odem, 2004), and new institutional demands (Rich and Miranda, 2005). Some of the most profound, but also most understudied, changes in such cities have taken place in their neighbourhoods, where Latino residents and long-term black and white residents increasingly live side by side.

The salience of such transitioning neighbourhoods in not only *immigrant* politics but also *urban* politics is particularly clear in Nashville, where a series of historically white, working-class neighbourhoods are now home to a growing Latino population. To demonstrate the place of these changing neighbourhoods in Nashville's wider urban politics and to show the ways that immigrant settlement complicates the relationship between neighbourhoods and the city, this chapter draws on ethnographic work in Nashville from 2006 to 2008. In 2006, approximately forty interviews were conducted with key actors involved with Nashville's neighbourhoods, a group including urban planners; Metro[2] officials; social-service providers; directors of non-profit organizations; and others charged with addressing Nashville's immigrant populations or its neighbourhoods. Interviews focused on how participants understood Nashville's immigrant population, when and where the city's immigrant groups impacted their work and organizations, and how immigrants in Nashville fit within broader understandings of and dynamics in the city.

These interviews were supplemented with archival work on Nashville's urban transformations since the late 1940s, review of media coverage of Nashville neighbourhoods since the mid-1990s, analysis of city documents concerning immigrants and neighbourhoods, as well as participant observation at urban planning events, city listening forums, and other venues. In 2007, additional interviews were conducted with long-term black and white residents and immigrants in the neighbourhoods under discussion. Interviews with residents focused on, among other things, how different neighbourhoods had been affected by immigrant settlement, where immigrant settlement fit within broader understandings of neighbourhood change, how immigrants themselves understood and adjusted to life in Nashville and its neighbourhoods, and how, if at all, they were involved with neighbourhood governance. In short, these interviews addressed how different groups came to grips with Nashville's emergence as a multicultural city in the 2000s.

The literature on the multicultural city and the politics of urban diversity is large and interdisciplinary (e.g. Flores with Benmayor, 1997; Nijman, 1997; Rocco, 1997; Kofman, 1998; Amin, 2002; Flores, 2003; Secor, 2004; Foner, 2007). Collectively, it foregrounds the transformative power of everyday encounters in shaping the politics of cultural belonging, while centring 'the perspective of citizens as social actors struggling not only to gain full membership in society but also to reshape it' (Flores 2003: 296). Works within urban geography have analysed urban diversity from the perspective of neighbourhoods (Martin, 2003b; Secor, 2003b), schools (Secor, 2004), workplaces (Secor, 2003a), and other sites, engaging 'the productive capacity of multiculturalism as lived through "the everyday"' (Clayton 2008: 255) and the ways that difference is drawn into and impacts urban politics.

The remainder of this chapter contributes to these discussions of the multicultural city by examining how Nashville in the 2000s saw and did not see 'diversity' in its transitioning neighbourhoods and how this seeing was connected to immigrants' place in Nashville's urban politics more broadly. Describing how Nashville neighbourhoods formalized their relationships to Metro government as the city's immigrant population grew, it argues that because Latino immigrants had an uneven presence in practices of neighbourhood governance, especially neighbourhood associations, they were largely invisible to the city *as neighbourhood residents* and absent from key parts of the city's urban politics. This situation, the chapter concludes, raises questions concerning how urban politics proceed in the multicultural city and who is institutionally allowed to be part of those politics. It also forces us to think more critically about how urban spaces are produced and claimed, as well as how we study them.[3]

How to make a neighbourhood

Lise Nelson and Nancy Hiemstra (2008) ask, 'how can scholars assess the changing nature of belonging and place within immigrant-receiving communities?' (322). One way to do so is to investigate how immigrants are, and are not, drawn into the governance of urban neighbourhoods, where the politics of immigrant integration and/or exclusion are often starkest. Understanding how immigrant settlement impacted the governance of Nashville neighbourhoods, however, first requires understanding the place of neighbourhoods themselves in Nashville's urban politics. In the Music City, neighbourhood involvement has roots in 1970s debates over interstate expansion (Daughtery, 1980), 1960s struggles toward desegregated schools (Woodward, 1978), and early twentieth-century urban political machines (Doyle, 1985). Since the late 1990s, neighbourhoods have played an especially important role in Nashville. In 1999, Bill Purcell was elected Nashville's mayor on a platform of neighbourhood empowerment that offered 'a neighbourhood-based approach to addressing the [city's] problems and opportunities' (Jones, 2000: 5). Shortly after taking office, Purcell made two decisions that impacted the place of neighbourhoods in Nashville's urban politics. First, he hired a new-urbanist planner, who made the neighbourhood the planning unit in Nashville. Under his leadership, planners began to see Nashville through human-scale neighbourhoods with streetscapes, limited traffic, and strong urban corridors. Although a focus on downtown development remained, urban planning increasingly included input from local residents and 'the neighbourhood', whose voice carried increasing weight in planning decisions and initiatives across the city.

Second, Purcell formed a Mayor's Office of Neighborhoods (MON) to act as liaison among his office, other Metro departments, and Nashville neighbourhoods. To provide information and technical assistance to residents, MON created the Neighborhood Training Institute, 'a series of workshops designed to build capacity and assist in the establishment and development of neighborhood associations' (*How to Be a Better Neighbor*, 2006: 19). Through these and other activities, MON played a key role in expanding neighbourhood associations across the city. When it began in 1999, approximately 125 neighborhood groups were identifiable in Nashville. By 2006, that number reached closer to 600; and neighbourhood empowerment was a buzzword across the city. MON, however, not only expanded the number of neighbourhood associations in existence but also elevated their importance as the institutionally visible representative of neighbourhoods and residents throughout the city. Under this model, when Nashville saw its residential spaces in the 2000s, it saw them through neighbourhood associations.

Luke Desforges and his co-authors (2005) suggest that in the current moment, 'active citizens are judged to have succeeded or failed ... as a place-based community, with repercussions for the further treatment of that locality by the state' (441). This link between active citizenship and official recognition of a locality was especially strong in Nashville. Through the confluence of the neighbourhood empowerment that Purcell touted and the wider neoliberal devolution of responsibility that Nashville and other cities were experiencing (Martin, 2003a, 2003b; Herbert, 2005), local citizenship practices became increasingly important in the Music City in the 2000s (Jones, 2000). At the turn of the twenty-first century, as Nashville experienced rapid economic and population growth (Winders, 2006b), its neighbourhoods began to organize and seek recognition from the city in response to pressures from downtown development, concerns about gentrification, and in parts of the city, the impacts of immigration.

Within Nashville, immigration's residential impacts are most visible in the southeast quadrant, where both Latino immigrants and the diverse groups of refugees brought to Nashville since the 1970s have settled (Winders, 2006a). Neighbourhoods in this area have been home to white, working-class communities with strong local identities since the 1920s. By the early 2000s, however, parts of southeast Nashville were experiencing turnover, as white residents aged and houses became available for purchase or rent, often by immigrants. In these neighbourhoods, the meeting of a greying white population and a growing Latino population led to rapid, and dramatic, change from owner-occupied to rented housing, from English to Spanish, and from elderly white residents who lived all their lives in the neighbourhood to young Latino families who lived their lives across international borders.

How did these neighbourhoods respond to such demographic, racial, and cultural change? How did Nashville as a city see and understand these residential transitions? Mike Davis (2000) notes that in new destinations, Latino residents often bring 'redemptive energies to the neglected, worn-out cores and inner suburbs' (51; see also Rocco, 1997), and this was certainly true in southeast Nashville. As the area's immigrant population grew, its business districts, which had been hard hit in the 1980s, began to thrive again. Key in the minds of many long-term residents was the 'overnight' emergence of car dealerships in southeast Nashville. In a city with limited public transportation and a sprawling urban landscape, owning a car was one of the only mobility options for immigrants. This need caught the attention of auto dealers, who, along with Mexican groceries, clothing stores, and Western Unions, transformed southeast Nashville's roads into the city's international corridors. Linked to immigrant settlement by long-term residents, such developments, especially car lots, came to represent a wider 'intrusion' against which residents had to 'hold our boundaries', in the words of one association president, to protect a sense of 'community'.

Under Nashville's model of neighbourhood empowerment, the ability to 'hold our boundaries' and preserve a sense of community lay largely in the hands of neighbourhood associations, which were the conduit through which zoning proposals and changes were partially vetted by Metro government. Residents and business owners desiring to change the status of their property or its structures, for example, presented their cases to neighbourhood associations, which then made recommendations to Nashville's Planning Department. Residents involved in neighbourhood associations, however, were empowered not only to influence zoning changes but also to police their own residential spaces. Through the Neighborhoods Organized to Initiate Code Enforcement (NOTICE) programme, Metro government trained residents to monitor and report codes violations in their neighbourhoods (*How to Be a Better Neighbor*, 2006). Codes violations, and the effort to convey codes standards to immigrants, were tense issues in southeast Nashville. For many long-term residents, concerns about their neighbourhood's changing material conditions entangled with questions about its cultural diversity, making codes violations a gauge of who was and was not an appropriate 'neighbour' (Shutika, 2005). Where (and how many) vehicles were parked at a dwelling, how many residents lived in one house, and whose responsibility it was to maintain rental units became points of contention and surveillance for long-term residents intent on maintaining a particular version of their neighbourhood's landscape even as its ethnic composition changed (Brettell and Nibbs, 2011). In the 2000s, as neighbourhood associations sprang into action across southeast Nashville, not only the ethnic and racial composition but also the politics of the neighbourhoods they represented became increasingly complex; and the realities of multicultural Nashville played out intensely on its residential streets. When that residential multicultural scene was translated to the scale of Nashville's wider urban politics, however, some of the detail, and some local residents, fell out of the picture.

How to mark the neighbourhood

Across southeast Nashville, many neighbourhood organizations predated both Mayor Purcell's focus on neighbourhoods and the arrival of Latino immigrants in the late 1990s. The combination of these two factors, however, added new motivations for neighbourhoods to formalize their associations, which, once officially recognized, could tap a range of services from Metro government. Equally important, as the most local form of political representation, neighbourhood associations formed the institutional link between the city and its residents. Metro funding for neighbourhood events was distributed through associations; and local actors, such as police officers and political candidates, reached and interacted with residents through neighbourhood associations. Being a formal neighbourhood association thus mattered in terms of what services and programmes were available *and* what local concerns gained visibility within Metro government (Herbert, 2005; Future of Neighborhoods, 2006). As the director of the Department of Codes and Building Safety explained, his department 'encourage[d] folks to work together within the frame of a neighborhood organization' through which residents 'get together, come to consensus, and decide, OK, what is our common problem, and then approach government to solve that problem'. In similar fashion, MON offered assistance 'to the [neighbourhood] groups that are there', linking MON's involvement to formal neighbourhood organizations. In this way, to speak to Metro government, and to speak within Nashville's urban politics *as neighbourhood residents*, individuals had to speak through the neighbourhood associations which represented where they lived.

The process of marking the limits of neighbourhood, however, was frequently contested (Martin, 2003a). On a basic level, residents had different understandings of neighbourhood (Campbell et al., 2009). As an urban planner in southeast Nashville noted, 'Even within the leadership of a group, you've got four different views of what the boundaries are.' Thus, forming a neighbourhood association was also deciding whose mental map of neighbourhood became dominant and whose was erased. To complicate things even further, once neighbourhood boundaries were determined, associations had to choose who, and whose issues, within them merited representation. In the context of southeast Nashville's growing racial and ethnic diversity, those decisions were increasingly difficult. As one non-profit director noted, 'while the neighborhood residents really want everybody to be welcomed, the fact is that it's really awkward when you are trying to mix a lot of different cultures'. Some neighbourhood associations attempted to reach immigrants by encouraging them to attend meetings, exploring the idea of advertising events in Spanish, and holding neighbourhood festivals; and some immigrant leaders tried to work with neighbourhood associations. Mostly, however, neighbourhood associations remained controlled by, and composed of, white residents (Shutika, 2005). In the words of one Metro official, neighbourhood associations 'don't have the participation of the immigrant population ... People are making decisions for those groups. They are not participating, but they are being discussed.'

This question of social inclusion became thornier when it was acknowledged that not all residents in a neighbourhood even recognized it as such (Herbert, 2005). While it is not unusual for residents to identify 'multiple and intersecting' boundaries in defining neighbourhood (Campbell et al., 2009: 463), for many Latino immigrants in southeast Nashville, 'neighbourhood', if it registered Nashville at all, included all Spanish speakers or individuals from a given town or country, rather than the bounded residential space identified as 'the neighbourhood' by long-term residents. As in southeast Los Angeles (Rocco, 1997), in southeast Nashville, cultural space partially trumped physical space as the basis of Latino community formation; and neighbourhood as a geographically bounded site carried less import for Latino immigrants than the pan-ethnic networks that stretched across Nashville's greater metropolitan area. When a well-known Hispanic organization ran community outreach, for example, it worked across Nashville, basing where it went on where members lived, rather than on a geographically defined neighbourhood. Along the same lines, a local community centre known for serving immigrants worked with Latino residents across Nashville and even nearby counties, rather than only with immigrants in the immediate neighbourhood where the centre was located.

In southeast Nashville, then, multiple understandings of 'community' and 'neighbourhood' spatially overlapped yet institutionally pulled apart, as immigrants created 'community' *across* neighbourhoods and long-term residents created associations to represent geographically discrete neighbourhoods. In Nashville's system of urban governance, neighbourhood associations were the scalar link between residents and the city. There was no institutional space for or recognition of immigrant social networks that stretched *across* neighbourhoods. This difference between ethnically defined immigrant *communities* and geographically bounded *neighbourhoods* raised knotty questions for neighbourhood associations tasked with representing increasingly diverse populations. It also complicated Nashville's efforts to empower local residents through the space and scale of the neighbourhood (Campbell et al., 2009). Immigrants living in southeast-Nashville neighbourhoods remained invisible to the city as neighbourhood residents because they were not present in the neighbourhood associations through which the city saw its residential spaces and because they generally did not recognize the neighbourhood as a space or scale through which they could make claims as city residents. Martin (2003b) suggests that 'for neighborhood-based organizations, place provides an important mobilizing discourse and identity for collective action, one that can obviate diverse facets of social identity in order to define a neighborhood-based polity' (730). In southeast Nashville, in the midst of demographic change, that process of mobilizing place to obviate difference through collective action was the goal of neighborhood associations. It was not always successful, however, because neighborhood associations faced the question of how to represent increasingly diverse urban populations, who understood neighbourhood/community in different ways, through a system that saw unified, discrete neighbourhoods.

How to make neighbourhood(s) matter

> [Diversity] increases my work because in the neighborhood where there is not as much diversity, you can figure out what the big picture is and so work on pieces of it ... In a neighborhood like this, there is *not necessarily one big picture*. Everybody's version of what the picture is, is

different, and my job is to help the residents attain what they think the big picture is, and so ... the biggest challenge is to sort through all this stuff I could be doing and figure out what are the things that I can help this group with, that ultimately will affect *the whole neighborhood*. (Director of a southeast-Nashville community organization)

The institutional realities of the multicultural neighbourhood and, by extension, city described by this director encapsulate the challenges that urban neighbourhoods like those in southeast Nashville faced as they shifted from addressing the complexities of daily life amidst rapid demographic change in the neighbourhood to making their concerns visible at 'a vantage point of 5000 feet', in the words of one Metro director. From this distance, Nashville saw its neighbourhoods *through* the organizations that represented them. In this move from the intimate scale of the neighbourhood to the more removed scale of the urban, the messy, contested spaces of the multicultural city in which residents understood and enacted 'neighbourhood' and 'community' in different ways became a mosaic of identifiable, bounded neighbourhoods – a visualization perhaps clearest in the map that lined the wall at MON and presented Nashville as a crazy quilt of contiguous neighbourhood associations. *Within* these neighbourhood spaces, the 'big picture' representing 'the whole neighbourhood' was not clear, as ethnic and racial diversification through immigrant settlement challenged the ability of neighbourhood associations, or even groups of neighbours, to speak for the neighbourhood as a whole. As these diverse neighbourhoods were (re)presented to the city, however, that 'big picture' had to be made clear by neighbourhood associations, if they wanted recognition from the city. The act of making a neighbourhood's 'big picture' visible within Nashville's urban politics, thus, was also the act of highlighting some aspects of that neighbourhood and occluding others.

This disjuncture between Metro Nashville's vision of its residential spaces and the realities of its multicultural neighbourhoods points to the limits of recent theorizations of the multicultural city (Table 9.1). Within geography, Ash Amin (2002) has described 'possibilities for urban interculturalism' (960) that stand in stark contrast to how Nashville, touted as a successful model of neighbourhood empowerment, saw its residential spaces. For Amin (2002), everyday shared sites, like those I have described in southeast Nashville, demand not 'the pursuit of a unitary sense of place', such as that supporting Nashville's neighbourhood empowerment, but instead, 'initiatives that exploit the potential for overlap and cross-fertilisation' (972). In his view, '(m)ixed neighbourhoods need to be accepted as the spatially open, culturally heterogeneous, and socially variegated spaces that they are, not imagined as future cohesive or integrated communities' (2002: 972). In these sites, Amin suggests, residents learn to live with diversity through 'constant negotiation, trial and error, and sustained effort' (976).

Table 9.1 The multicultural city

	Main argument	View of space	Perspective on city
Amin's urban interculturalism	Urban spaces are composed of 'multiple publics' that are culturally heterogeneous.	Space and social acceptance are produced through cultural negotiation, contact, and sometimes conflict.	The city is seen from the scale of everyday interactions that drive the 'multiple publics' that comprise urban interculturalism.
Fainstein's just city	Urban spaces should be addressed through the lens of a 'collective good' that does not privilege difference.	Space is produced through efforts to plan for a just and equitable city for all.	The city is seen as a whole through the public realm that should shape planning efforts.
Flores's Latino cultural citizenship	Disenfranchised groups carve out distinct cultural and social spaces from dominant society.	Cultural and social space is produced through a zero-sums game of claiming space from another group.	The city is seen as a site of spatial struggle between dominant and disenfranchised groups.

Urban scholar Susan Fainstein (2005) takes a different approach. In a reflection on what diversity means in urban planning and how, and whether, it is an attainable goal, she argues for attention to a 'collective identity that does not privilege difference over other goods' (14). Fainstein is, thus, less sanguine about how cities institutionally address diversity and cautions against Amin's optimistic assessment of sites that support 'multiple publics' (2002: 972), by focusing on how they play out in practice. For Fainstein, efforts to bring notions of diversity into urban planning should not stand in the way of efforts to create a just city and the political consciousness needed to sustain it. Instead, 'diversity', as both a selling point and reality of multicultural cities, may not necessarily feed into equity and the 'broadly satisfying public realm' (2005: 9) whose differences Amin positions as central to the multicultural city and whose shared goals Fainstein sees as central to the just city. Where Amin locates political possibility in communities that are not necessarily cohesive, Fainstein worries about broader efforts to create social justice through them.

Both Amin's and Fainstein's arguments diverge from William Flores's (2003) description of how Latino communities respond to their place in multicultural cities. For Flores, Latino men and women must struggle 'for a distinct social space in which members of the marginalized group are free to express themselves and feel at home' (297). In this way, he suggests, Latino/as, as a disenfranchised group, can only find their place in the multicultural city by carving out a distinct space *from* the spaces of dominant society. The multicultural city that holds possibilities for Amin and that presents challenges for Fainstein, then, is, for Flores, a site of struggle in which space is a zero-sums game where one groups (re)claims it from another. To belong in the multicultural city, Flores argues, marginalized groups like Latino/as must mark out their own social space defined against and separate from the dominant society.

The differences among these three authors' reflections on how the multicultural city functions and how different groups within it interact demonstrate the need for analyses that see the city, its spaces, and its residents from different vantage points and that start with a more nuanced understanding of urban space. For Amin, the multicultural city begins with its everyday spaces and neighbourhoods, which can be celebrated for their contradictions, rather than managed for a greater good, as Fainstein would suggest. Amin's approach, however, does not engage the infrastructures of power and unequal access that contour Fainstein's view of the city and that, in the case of Nashville, rendered immigrant claims to place in the neighbourhood largely invisible within a system of governance organized around neighbourhood associations. For Fainstein, by contrast, the multicultural city begins with planning imperatives caught between the recognition of diversity and the quest for equity across the city. Both ends of the spectrum that Fainstein lays out between diversity and equity, however, are divorced from the lived realities that Amin places at the centre of his multicultural city and, instead, come into view at a spatial remove from urban neighbourhoods – at the view from 5000 feet. At this remove, not all residents' struggles to make place or community are visible or acknowledged. Finally, Flores's vision of the multicultural city begins with dominant spaces that structure the city and exclude Latino residents, who must claim back space and the right to be present. His 'distinct social space' of Latino cultural citizenship (Flores with Benmayor, 1997: 1), though, cannot account for Amin's urban sites that support 'multiple publics' or Fainstein's inclusive 'public realm', as it is based on a winner-takes-all understanding of urban space in which one group has it and another does not. To claim cultural citizenship in the city, Latino/as, for Flores, must claim space back from the dominant group, not occupy it together and, in the process, transform its meanings. Although based in the multicultural city, Flores's Latino cultural citizenship cannot account for multicultural, *shared* spaces.

In the 2000s, Nashville was caught among these approaches to the multicultural city, wanting, like Amin, to acknowledge diversity among residents, trying, like Fainstein, to ensure representation for all neighbourhoods, and working, like Flores, to allow groups to claim place by centring the neighbourhood association as a link between residents and the city. This structure of urban governance, however, even as it called for neighbourhood empowerment, was limited in its ability to do so *because of* the ways that immigrant settlement changed Nashville neighbourhoods and complicated the link between the 'public' and the institutions meant to represent it. Because Nashville was locked in a spatial ontology of discrete neighbourhoods represented by unified neighbourhood associations,

Amin's spaces of 'multiple publics' and Flores's transgressive sites of Latino cultural citizenship were invisible in a system where one neighbourhood association represented a group of presumably uniform residents in a discrete residential space. Because that spatial ontology did not acknowledge difference *within* neighbourhood spaces, or communities *across* them, Fainstein's 'broadly satisfying public realm' was also only partially realizable, produced, as it is, across urban spaces. In the context of multicultural Nashville, Latino acts of claiming space in the neighbourhood were rendered institutionally invisible, if inadvertently, by long-term residents working through neighbourhood associations to make their own place in the city visible to Metro government. Through the institutional map that envisioned Nashville as a series of bounded neighbourhoods represented by neighbourhood associations, the city could not see immigrant communities that lived *across* southeast Nashville. Although the city was aware of immigrants through its police department, school system, and other institutions, as well as through public discourse surrounding immigrant legality and crime, within its neighbourhoods, immigrants were hard to find because of how the city saw those spaces. As a result, within Nashville's urban politics, within the struggles 'that govern the use of space and place' (Pierce et al., 2011: 55), immigrants *as neighbourhood residents* were largely absent.

Nashville's way of seeing and engaging its neighbourhoods, then, had material consequences for immigrant inclusion. The same neighbourhood associations that struggled to include immigrants in their activities *in the neighbourhood* increasingly assumed responsibility for representing immigrants as part of that neighbourhood *to the city at large* within Nashville's governance. As has been well documented, immigrants in new destinations are often 'not empowered to claim belonging locally' (Nelson and Hiemstra, 2008: 329) and face a series of cultural challenges to claiming place in their new homes. As I have shown here, however, some of those challenges are also institutional. Nashville neighbourhoods in the 2000s, at the scale of everyday interaction, resonated with Amin's urban interculturalism. Within these neighbourhoods, immigrant place-making occurred in sites already imbued with social and cultural meaning for long-term residents, making them spaces of cultural exchange *and* making practices of Latino cultural citizenship processes of claiming space with, alongside, and against long-term residents' own place-making efforts, rather than claiming space from them, as Flores argues. At the scale of the overall city, however, those neighbourhoods *as* sites of intercultural exchange appeared as uniform neighbourhoods, not multicultural sites, as the city tried to see all residents in an equitable way through neighbourhood associations.

In Nashville, when Latino residents will claim membership within 'the imagined urban community' (Secor, 2003b: 164) – whether through its neighbourhood or elsewhere in the urban fabric – is unclear. Part of that effort to claim membership in the local urban community, however, will be bound up with immigrants' abilities to claim an *institutional presence* within Nashville's urban politics as neighbourhood residents who merit attention from the city alongside long-term residents. For immigrants to achieve this institutional presence, cities like Nashville will have to learn to see and interact with not only their 'community of neighbourhoods' but also the multiple publics and communities that those neighbourhoods support. They will also have to learn to see across the scales of urban politics and the different roles that neighbourhoods play in those politics. If we, as scholars, are to understand the realities of contemporary multicultural cities like Nashville, we need a theorization of urban space that can address these different ways that neighbourhoods and urban politics are shaped, contested, and linked. We also need methodological and conceptual frameworks that can engage multiple ways of seeing neighbourhoods themselves and their place in wider urban politics. To understand Amin's urban interculturalism, Fainstein's just city, Flores's Latino cultural citizenship, and the relationship among them, we must approach urban space as 'full of internal conflict' (Massey, 1994: 155) and multiple, as Nashville's transformation into a multicultural city makes clear.

Notes

1 The author thanks Deb Martin and Mark Davidson for their insightful feedback on earlier drafts of this chapter. The Russell Sage Foundation funded both the study on which the chapter is based and the time during which I wrote it as a visiting scholar. I am thankful for both opportunities.
2 Nashville has had a combined city–county Metropolitan government since 1963 (Coomer and Tyler, 1974).
3 For an extended discussion of these ideas, see Winders 2013.

References

Amin, A. (2002) 'Ethnicity and the multicultural city: Living with diversity', *Environment and Planning A*, 34(6): 959–980.
Anderson, K. (1987) 'The idea of Chinatown: The power of place and institutional practice in the making of a racial category', *Annals of the Association of American Geographers*, 77(4): 580–598.
Brettell, C. and Nibbs, F. (2011) 'Immigrant suburban settlement and the "threat" to middle class status and identity: The case of Farmers Branch, Texas', *International Migration*, 49(1): 1–30.
Brubaker, R. (2001) 'The return of assimilation? Changing perspectives on immigration and its sequel in France, Germany, and the United States', *Ethnic and Racial Studies*, 24(4): 531–548.

Campbell, E., Henly, J., Elliott, D., and Irwin, K. (2009) 'Subjective constructions of neighborhood boundaries: Lessons from a qualitative study of four neighborhoods', *Journal of Urban Affairs*, 31(4): 461–490.

Clayton, J. (2008) 'Everyday geographies of marginality and encounter in the multicultural city', in Dwyer, C. and Bressey, C. (eds) *New Geographies of Race and Racism*. Ashgate: Aldershot. pp. 255–267

Coomer, James and Tyler, Charlie (1974) *Nashville Metropolitan Government: The first decade*. Bureau of Public Administration. University of Tennessee.

Daugherty, Tracy (1980) *Community Involvement Shapes a Highway: The redesign of Nashville's I-440*. Environmental Action Plan Report. Number 10. Federal Highway Administration: U.S. Department of Transportation.

Davis, Mike (2000) *Magical Urbanism: Latinos reinvent the US city*. Verso: London.

Desforges, L., Jones, R., and Woods, M. (2005) 'New geographies of citizenship', *Citizenship Studies*, 9(5): 439–451.

Doyle, Don (1985) *Nashville since the 1920s*. University of Tennessee Press: Knoxville, TN.

Ellis, M. and Wright, R. (1998) 'The balkanization metaphor in the analysis of US immigration', *Annals of the Association of American Geographers*, 88(4): 686–698.

Fainstein, S. (2005) 'Cities and diversity: Should we want it? Can we plan for it?', *Urban Affairs Review*, 41(1): 3–19.

Flores, W. (2003) 'New citizens, new rights: Undocumented immigrants and Latino cultural citizenship', *Latin American Perspectives*, 129(30): 295–308.

Flores, W. with R. Benmayor (1997) 'Constructing cultural citizenship', in Flores, W. and Benmayor, R. (eds) *Latino Cultural Citizenship: Claiming identity, space, and rights*. Beacon Press: Boston. pp. 1–23.

Foner, N. (2007) 'How exceptional is New York? Migration and multiculturalism in the Empire City', *Ethnic and Racial Studies*, 30(6): 999–1023.

The Future of Neighborhoods: A vision for the future of Nashville and Davidson County, Tennessee. (2006) Neighborhood Resource Center for the Neighborhoods of Nashville.

Herbert, S. (2005) 'The trapdoor of community', *Annals of the Association of American Geographers*, 95(4): 850–865.

Heynen, N. (2009) 'Bending the bars of empire from every ghetto for survival: The Black Panther Party's radical antihunger politics of social reproduction and scale', *Annals of the Association of American Geographers*, 99(2): 406–422.

How to be a better neighbor (2006) Metropolitan Government of Nashville and Davidson County. Department of Codes Administration.

Jackson, Mandi (2008) *Model City Blues: Urban space and organized resistance in New Haven*. Temple University Press: Philadelphia, PA.

Jones, Grant (2000) *Developing a Neighborhood-focused Agenda: Tools for cities getting started*. Annie E. Casey Foundation: Baltimore, MD.

Kofman, E. (1998) 'Whose city? Gender, class, and immigrants in globalizing European cities', in Fincher, R. and Jacobs, J. (eds) *Cities of Difference*. Guilford Press: New York. pp. 279–300.

Ley, D. (1995) 'Between Europe and Asia: The case of the missing sequoias', *Ecumene*, 2(2): 185–210.

Martin, D. (2003a) 'Enacting neighborhood', *Urban Geography*, 24(5): 361–385.

Martin, D. (2003b) '"Place-framing" as place-making: Constituting a neighborhood for organizing and activism', *Annals of the Association of American Geographers*, 93(3): 730–750.

Massey, Doreen (1994) *Space, Place, and Gender*. University of Minnesota Press: Minneapolis: Minneapolis, MN.

Massey, Doreen (ed.) (2008) *New Faces in New Places: The changing geography of American immigration*. Russell Sage Foundation: New York.

Massey, Douglas and Denton, Nancy (1993) *American Apartheid: Segregation and the making of the underclass*. Harvard University Press: Cambridge, MA.

McClain, P., Carter, N., DeFrancesco, V., Lyle, M., Nunnally, S., Scotto, T., Kendrick, A., Grynaviski, J., Lackey, G., and Cotton, K. (2006) 'Racial distancing in a southern city: Latino immigrants' views of black Americans', *Journal of Politics*, 68(3): 571–584.

Nelson, L. and Hiemstra, N. (2008) 'Latino immigrants and the renegotiation of place and belonging in small town America', *Social and Cultural Geography*, 9(3): 319–342.

Nijman, J. (1997) 'Globalization to a Latin beat: The Miami growth machine', *Annals of the American Academy of Political and Social Science*, 551: 164–178.

Odem, M. (2004) 'Our Lady of Guadalupe in the New South: Latino immigrants and the politics of integration in the Catholic church', *Journal of American Ethnic History*, 24(1): 26–57.

Park, Robert, Burgess, Ernest, and McKenzie, Roderick (1925) *The City*. University of Chicago Press: Chicago.

Peach, C. (1996) 'Good segregation, bad segregation', *Planning Perspectives*, 11(4): 379–398.

Pierce, J., Martin, D., and Murphy, J. (2011) 'Relational place-making: The networked politics of place', *Transactions of the Institute of British Geographers*, 36(1): 54–70.

Rich, B. and Miranda, M. (2005) 'The sociopolitical dynamics of Mexican immigration in Lexington, Kentucky, 1997 to 2002: An ambivalent community responds', in Zúñiga, V. and Hernández-León, R. (eds) *New Destinations: Mexican immigration in the United States*. Russell Sage Foundation. New York. pp. 187–219.

Rocco, R. (1997) 'Citizenship, culture, and community: Restructuring in southeast Los Angeles', in Flores, W. and Benmayor, R. (eds) *Latino Cultural Citizenship: Claiming identity, space, and rights*. Beacon Press: Boston. pp. 97–123.

Secor, A. (2003a) 'Belaboring gender: The spatial practice of work and the politics of "making do" in Istanbul', *Environment and Planning A*, 35(12): 2209–2227.

Secor, A. (2003b) 'Citizenship in the city: Identity, community, and rights among women migrants to Istanbul', *Urban Geography*, 24(2): 147–168.

Secor, A. (2004) '"There is an Istanbul that belongs to me": Citizenship, space, and identity in the city', *Annals of the Association of American Geographers*, 94(2): 352–368.

Shutika, D. (2005) 'Bridging the community: Nativism, activism, and the politics of inclusion in a Mexican settlement in Pennsylvania', in Zúñiga, V. and Hernández-León, R. (eds) *New Destinations: Mexican immigration in the United States*. Russell Sage Foundation: New York. pp. 103–132.

Winders, J. (2005) 'Changing politics of race and region: Latino migration to the U.S. South', *Progress in Human Geography*, 29(6): 683–699.

Winders, J. (2006a) '"New Americans" in a "New South" city? Immigrant and refugee politics in the Music City', *Social and Cultural Geography*, 7(3): 421–435.

Winders, J. (2006b) 'Placing Latinos in the Music City: Latino migration and urban politics in Nashville, Tennessee', in Smith, H. and Furuseth, O. (eds) *Latinos in the New South: Transformations of place*. Ashgate: Aldershot. pp. 167–190.

Winders, J. (2008) 'An "incomplete" picture? Race, Latino migration, and urban politics in Nashville, Tennessee', *Urban Geography*, 29(3): 246–263.

Winders, J. (2013) *Nashville in the New Millennium. Immigrant Settlement, Urban Transformation, and Social Belonging*. Russell Sage Foundation: New York.

Woodward, David (1978) 'Busing plans, media agendas and patterns of white flight: Nashville, Tennessee and Louisville, Kentucky', PhD dissertation in Political Science, Vanderbilt University.

Wright, R., Ellis, M., and Parks, V. (2005) 'Re-placing whiteness in spatial assimilation research', *City and Community*, 4(2): 111–135.

Yarbrough, R. (2003) 'Latino/white and Latino/black segregation in the southeastern United States: Findings from Census 2000', *Southeastern Geographer*, 43(2): 235–248.

10
THE EMBEDDED POLITICS OF ENTREPRENEURS

Susan Hanson

Introduction

The question of where politics occurs in the city is centrally complicated by what constitutes 'politics'. A basic premise of this book is that politics appears in many guises other than the collective decision making of electoral activities and takes place in many forms and locations outside government and political parties. As informed citizens we must reflect on what we look for in urban politics and must seek out unexpected viewpoints. In this chapter I look at the everyday activities of entrepreneurs, most of whose businesses are small, as potential sites of urban politics. Although popular stereotypes of entrepreneurs portray them as independent, individualistic, self-reliant, and autonomous, entrepreneurs are the product of – and very much embedded in – social networks, communities, and places. I explore how this embeddedness is linked to a range of political actions in which business owners foster new forms of inclusion and empowerment among the least powerful members of their communities.

The chapter draws on data from in-depth interviews with 200 business owners in Worcester, Massachusetts and 180 in Colorado Springs, Colorado. Before examining the embedded politics of entrepreneurs, I shall look briefly at conventional thinking about entrepreneurship and urban politics, and shall consider how feminist thinking has made embedded politics visible through changes in understanding of what counts as politics.

Entrepreneurs, politics, and the local state

The primary source of geographers' interest in entrepreneurial processes has been the presumed central role of entrepreneurs in generating economic growth (Malecki, 1994, 1997). Put another way, the main focus of research on the relationship between entrepreneurs and the places where their businesses are located has been on entrepreneurship as a driver of economic growth and hence on the place characteristics that foster entrepreneurship. To be sure, one of the main functions of the set of institutions and actors comprising the local state is to foster economic growth so as to ensure the tax base needed to support the governing activities of the state (Tilly, 1990).

This central interest in entrepreneurs as growth (and job) creators has also coloured definitions of entrepreneurs (e.g. as people who launch innovative firms that grow quickly) and has led some observers to withhold the esteemed moniker *entrepreneur* from those business owners with few or no employees and with stable but not growing revenues. This latter category of business owners has been referred to as the 'merely self-employed' to highlight their lack of contribution to the dynamic local economies that are deemed attributable to 'true entrepreneurs' (Aronson, 1991). To be labelled an entrepreneur, does your business need to achieve a certain size and measure of impact? Must it be deemed sufficiently innovative? As each of these characteristics is difficult to measure, most analysts have settled on using 'entrepreneur' to signal someone who owns a business, deals with the risks and uncertainties associated with ownership, and takes responsibility for the day-to-day operation of the business (Moore, 1990). This is the definition I have adopted, and as a result I use the terms *entrepreneur* and *business owner* interchangeably.

The emphasis on the economic impacts of entrepreneurialism has had several consequences for thinking about entrepreneurship, politics, and the state. First, it directs attention to the politics surrounding the formal policies of the local state aimed at attracting job- and revenue-producing firms from elsewhere to locate in one local jurisdiction instead of another (Malecki, 1994). Second, it suggests that entrepreneurs' major form of political involvement will be directly related to promoting the growth of their firms and therefore involve serving on elected bodies such as the City Council or entail interactions with formal institutions of the local state such as local economic development agencies. Third, in terms of the location of political activity, it emphasizes the formal spaces of interaction of entrepreneurs and officials of local government. Fourth, it suggests that entrepreneurs' main contributions to the well-being of the people and places in which they live are limited to their firms' roles as employers of labour and sources of taxes. And finally, it signals that the only business owners who matter politically to the well-being of a place are those who launch and lead high growth, and preferably large, firms.

The focus on the entrepreneurialism – growth nexus ignores the variety of informal ways in which entrepreneurs affect places, discounts entrepreneurs' political activities that are not directly connected with economic growth, and overlooks the importance of personal relationships as a site of political activity. The empirical portion of this chapter shifts attention to these neglected politics of entrepreneurs. My understanding of the embedded politics of entrepreneurs owes a great deal to feminist insights about what 'politics' is all about.

Feminist contributions to understanding politics

Feminists have been instrumental in expanding the meaning of politics beyond its conventional understanding as the means of collective decision making, usually within government structures. Any feminist consideration of politics comes out of a feminist tradition of wanting to understand the processes and everyday actions through which the meanings of taken-for-granted words and categories, such as *gender* (West and Zimmerman, 1991) or *experience* (Scott, 1992), are created and recreated in certain contexts. According to Patricia Martin (2004: 17), feminists have seen politics present 'wherever power is perceived to exist'. Feminist political geographer Eleonore Kofman's (2005: 519) definition of politics focuses centrally on power: 'politics is about how people exercise power through both material and discursive practices (Peake, 1999) within specific places and across specific scales of analysis (Kodras, 1999).'

This focus on power draws attention to social relationships of all kinds and to the connections between people that such relationships define. In particular, politics is not confined to the actions of elite actors, but infuses the everyday actions of all people (Staeheli et al., 2004: 5–6).[1] If you agree with Megan Cope (2004: 75) that the goal of politics is, in part, 'to raise awareness, to change social or cultural practices, [or] to challenge the status quo', then you can imagine literally thousands of ways that people can act politically in the course of their daily lives. In each of these political acts the agent is involved in changing power relations. One such example would be a woman engineer supervising an all-male work unit; another would be a volunteer helping someone learn how to read. Note that the extent to which such acts change power relations or challenge the status quo depends on the geographic context in which they occur.

Martin et al. (2007: 79) have argued that politics is constituted through the everyday lives of individuals embedded in communities. Using their definition of activism as 'everyday actions by individuals that foster new social networks or power dynamics', they described several instances in which an entrepreneur's actions created new connections between people and shifted power relations within the community. While the focus in this chapter is on politics rather than activism per se, the analysis that follows extends the previous work in several ways, notably by considering a far wider range of entrepreneurs' actions and in linking those actions explicitly to the entrepreneur's business activities.

Data

The data come from a study aimed at understanding the relationships among gender, geography, and entrepreneurship. We conducted in-depth interviews with owners of 200 businesses in Worcester, Massachusetts in 1998–99 and 180 Colorado Springs, Colorado in 2000. These two places were selected because while similar in size (both had populations of about 400,000 in 1990 and both over 500,000 in 2000), but they differ in their business cultures. The differences are perhaps best captured in the phases Rust Belt city (Worcester) and Sun Belt city (Colorado Springs); although the analysis to follow does not emphasize the divergence between these two cities, some of the fundamental bases of difference will become evident.

Because they were randomly selected from a purchased list of privately owned businesses, the enterprises in the sample are extremely diverse, ranging across all industrial categories (agriculture; manufacturing; transportation, communications, utilities; wholesale trade; retail trade; finance, insurance, and real estate; service industries) and ranging from single-person operations to companies with several hundred employees. The sample includes businesses owned by women (45 per cent of the Worcester businesses; 42 per cent of the Colorado Springs businesses), by men (42 per cent of the Worcester sample; 40 per cent of the Colorado Springs sample), and by both women and men (13 per cent and 19 per cent for Worcester and Colorado Springs, respectively).[2] In addition to the interviews, a mailed survey was sent to a random sample drawn from the same list of businesses; in Worcester we received 340 usable returns; in Colorado Springs that number was 504. For a more detailed description of the research design see (Hanson, 2003).

Entrepreneurs' relationship to place

In view of the need for place-based knowledge to launch and operate a viable business, perhaps it is not surprising that entrepreneurs have lived longer in a place than have members of the general population (Hanson, 2003). One measure of length of residence in a place comes from the decennial census; it indicates if the person was living in the same metro area five years previously. We asked business owners how long they had lived in the metro area, and their responses show that a higher proportion of the entrepreneurs than of the general population have lived in the metro area at least five years: In Worcester, 84 per cent of the population were living in the metro area five years previously (1990 Census), but this was true for fully 97 per cent of the Worcester business owners who completed the mailed survey. In Colorado Springs the comparable figures were 64 per cent of the population and 93 per cent of the business owners. This single piece of evidence suggests that entrepreneurs have deeper roots in a place than those who do not own and run businesses there. These figures also indicate a difference in rootedness to place between Worcester and Colorado Springs.

In fact, the interviews shed additional light on the nature of entrepreneurs' relationship to place: Whereas 63 per cent of the Worcester business owners we interviewed had lived in the Worcester metro area their whole lives or almost their whole lives, only 12 per cent of the Colorado Springs respondents had been there that long. As is the case for most Sun Belt cities, Colorado Springs attracts more in-migrants – whether or not they end up owning businesses – than does Worcester; on the other hand, people who live in Worcester tend to stay there.

Why should we care about these measures of rootedness to place? Length of residence in a place, defined here as a metro area, can be viewed as an indicator of degree of embeddedness in that place or, put another way, as a measure of the number and nature of connections that an individual has developed with other people and with the organizations, cultures, and institutions there. Because studies have shown that inertia in place increases the longer someone lives there (Clark, 1996), length of residence might also be taken as a gauge of a person's level of commitment to place and therefore one's willingness to invest time and effort to create positive change there. It is clear from our interview data that the majority of business owners are committed to their home metro areas: 81 per cent of the entrepreneurs in Worcester and 70 per cent of those in Colorado Springs said that they had not considered starting their business in another metro area. Moreover, few (11 per cent in Worcester and 19 per cent in Colorado Springs) said they would consider moving their business elsewhere.

These indicators of entrepreneurs' ties to place suggest that they have well-developed social capital, a term Bourdieu (1985: 248) used to refer to 'the actual or potential resources which are linked to possession of a durable network of more or less institutionalized relationships of mutual acquaintance or recognition'. In other words, individuals, groups, and places can benefit from participating in networks of social relationships.

Entrepreneurs and the state

Before turning to examine the ways in which entrepreneurs' relationship to place affects the nature and spaces of their (mainly informal) engagement with urban politics, it is worth briefly considering the nature of the formal or official interactions through which the business owners in this study encounter the state. Government policies and programmes, promulgated by the state, are the outcomes of the collective decision making associated with the conventional definition of politics. In the interviews, we posed a series of questions about the entrepreneur's interactions with government; in particular, we asked if any government programmes had helped the entrepreneur in the start-up process, if any government programmes had hindered that process, and if the entrepreneur had made use of a business counselling service before starting the business. Answers to these questions indicate limited contact with the state before or during business start-up. They also indicate considerable confusion about what constitutes a government programme or service.

Almost all of the business owners we talked with (96 per cent of those in Worcester and 92 per cent of those in Colorado Springs) reported that no government policies or programmes had helped in starting the business, and roughly one-quarter (27 per cent in each place) said that government policies or programmes had hindered them. Taxes and a variety of specific regulations were the hindrances mentioned. Moreover, more than three-quarters of the entrepreneurs in each metro area said that they had not used a business counselling service before launching the business, although women were much more likely than men to have sought help from such a service: in Worcester 29 per cent of women vs. 12 per cent of men and in Colorado Springs 35 per cent of women vs. 17 per cent of men said they had used a business counselling service. The services mentioned were the Small Business Administration (SBA), which provides advice and loans, and SCORE, an organization of volunteer retired executives who provide advice and mentoring to new business owners; while SCORE partners with the SBA, it is a non-profit, non-governmental organization and not part of the SBA.

It was clear from the interview discussions on this topic that many entrepreneurs thought SCORE is a government programme and others did not realize that the SBA is a federal programme. This confusion suggests that the answers to the questions about help from government programmes should be viewed cautiously. To put this confusion in a larger context, Krugman (2012) reports that research by Suzanne Mettler has made startlingly clear that large numbers of Americans who benefit from federal programmes evidently do not know that government is the source of their benefits. For example, 44 per cent of Social Security recipients, 43 per cent of those receiving unemployment benefits, and 40 per cent of those on Medicare say that they 'have not used a government social program'.

In our interviews, this confusion was particularly intense around this question posed to women business owners: 'Is your business registered with the state of [Massachusetts/Colorado] as a woman-owned business?' In Worcester only 12 per cent of the owners of woman-owned businesses said they were registered as such with the state, whereas the comparable figure in Colorado Springs was 23 per cent. Typical responses were:

> I don't know. I am registered with the state of Colorado because I pay state income tax and all those good things.

> I can't tell you that; I don't know. I guess I'm probably not, I've never even thought about it. Where do you find out?

In fact, formal registration as a woman-owned business (WOB) is an elaborate process – and a memorable one because of the amount of paperwork involved. The business owners who had completed that paperwork described the process in detail, emphasizing how time-consuming it was; they also said that being registered as a WOB had not helped them very much.[3] Much of the confusion around WOB registration no doubt stems from the fact that all businesses must register with the local authority; hence we heard responses such as these to the question we posed about WOB registration:

> They know I'm a woman, and I'm registered, so hopefully [I'm registered as a WOB].

> It is a woman owned business, and it is registered.

Therefore, the actual percentages of woman-owned businesses formally registered as WOBs in these two metro areas may well be lower than 12 per cent and 23 per cent.

In addition to registering the business when they launch it, all entrepreneurs also encounter the local state through zoning regulations. Entrepreneurs operating home-based businesses, in particular, can find that local zoning ordinances pose barriers. A couple in Colorado Springs who ran a home-based small-run digital printing company mused:

> Our home owners' covenants prohibit this [home-based business] and until we move out, if anybody jumps on me it's going to be a home office because in this neighbourhood probably every other home has a home office in it. So far my fingers are crossed; my immediate neighbours know and they are cheering me on. It's not like I'm selling burgers or I've got six cars out in the yard that I'm working on.

When we asked, 'Is there was anything that local government could do to help your business?', the owner of a drive-through ice cream and sandwich shop told us about the battle he was having with the city's zoning rules about signs.

> The sign out front has been there for 40 something years ... but umm their argument was that ... I fixed up the sign on the building and now they say I have two signs that they consider signposts or billboard signposts ... We haven't completely resolved it yet; we're working on it.

By far the most frequent response to this question about how local government could help was to 'lower the taxes'.

> Just lower the taxes. Anything – it doesn't matter what kind of tax it is; it's tough as a small business owner ... We need less government, that's for sure. That's why you won't find too many business owners that aren't Republicans. And I know you being a school teacher, you're a Democrat for sure. I've got a lot of friends who are school teachers. We argue all the time. (Woman owner of a tavern; Colorado Springs)

In contrast to the frequent call, especially in Colorado Springs, for lower taxes and 'less government', a few entrepreneurs in the study had served as elected representatives in local government. The co-owner of an independent bookstore in Colorado Springs linked this form of direct civic involvement specifically to the embeddedness of his business in the community and to the kind of business he and his wife ran.

> We are convinced that booksellers have an obligation to their community much like a library does. There are great similarities. A bookseller must feel responsible for his or her community, and must be active, get out those front doors and do something. That's why she [his wife and co-owner of the store] is on the city council. We have a nine-member city council, and two of the members are booksellers ... She is, more than any other single individual, responsible for all this downtown innovation that's taking place. [He had also served on the city council.]

In sum, for many entrepreneurs, encounters with the state are infrequent and intermittent and affect the daily running of a business mainly through taxation. Few firms in any locale are large or fast-growing enough to enter into the politics of growth with local authorities. Many businesses do routinely deal with state and federal government regulations such as those covering the handling of toxic chemicals (e.g. for dry cleaners or furniture manufacturers) or the treatment of employees. Even business owners who strongly agree with the need for such regulations resent the paperwork and 'red tape' entailed. Some business owners view the state in adversarial terms rather than as a collaborator in creating a positive environment for living and running a business. Certainly, mention of 'government' did not stimulate thoughts about the relationship of the entrepreneur's business to place or community. Few spoke of government as a force for positive change in the community, and only three entrepreneurs in each metro area sample had served in elected positions in local government. In contrast, a look into some unexpected spaces of urban politics reveals myriad ways in which entrepreneurs engage in political action in their communities.

Entrepreneurship and urban spaces of politics

To what extent do entrepreneurs' high level of rootedness in, and assumed commitment to, the places in which they live and run their businesses translate into specific actions aimed at improving the quality of life in these places? Insofar as political action can be aimed at raising awareness, changing social or cultural practices, challenging existing policies (Cope, 2004: 75), actions aimed at changing the status quo by building the human capital of the most vulnerable people in an area or by caring for the environment can be seen as political. In their roles as citizens and as business owners, entrepreneurs initiate such actions in abundance.

In the interviews we asked specifically about the business owner's involvement in volunteer activities. Although many respondents evidently interpreted the question as referring to volunteering through formal organizations, the responses revealed the great diversity of ways in which entrepreneurs contribute to the life of their community. One theme is clear in all of the stories about business owners' contributions to community life: these individuals become involved in many ways because of their deep knowledge of community needs, knowledge that in turn is linked to their embeddedness in the community and for many entrepreneurs is tied specifically to the type of business they run.

One could argue that any involvement of a business owner in civic affairs is motivated by self-interest in that looking after the health/well-being of the community – however the business owner defines it – is likely to raise local awareness of, and yield some benefit to, his or her business. The line between community involvement to promote business interests versus such involvement to give voice to those who have been marginalized can be fuzzy. Almost always the form of an entrepreneur's community involvement emerges from their knowledge of, and their personal contacts within, the local community. Certainly the beneficiaries and impact of their involvement are local.

The analysis that follows no doubt describes volunteer activities that span a motivational continuum from high to low levels of self-interest. It is worth noting, however, that the interview data show no relationship between the proportion of an entrepreneur's sales that are local (a measure of how dependent the enterprise is on the local market) and volunteering. The average of the percentages of business sales that are local is roughly 82 per cent in Worcester and 72 per cent in Colorado Springs whether or not the owner volunteers. So those whose businesses rely heavily on the local market are no more likely to be actively involved in the community. Also significant is that entrepreneurs volunteer at about double the rate of the general citizenry: 54 per cent of the Worcester sample and 46 per cent of the Colorado Springs sample said they volunteered, compared with a rate of about 25 per cent for the population as a whole (Brady, 2002).

Many entrepreneurs contribute to the life of the community and are politically active citizens, but the interviews demonstrate how business ownership can add another dimension to citizenship. As *parents*, entrepreneurs volunteer in the schools, they coach youth sports, and they organize scouting activities; they may continue in these forms of volunteering long after their own children are no longer directly involved in them. As *citizens*, entrepreneurs volunteer for the American Red Cross, raise money for the American Cancer Society or the local hospital; they work in the community through their churches and volunteer fire departments, and they participate in the activities of organizations like Lions or Rotary clubs. Although such activities obviously contribute to community life in important ways, I do not consider them in further detail here because they are not directly connected to the entrepreneurs' businesses. Instead, I focus on entrepreneurs' community-oriented activities that do directly link to their businesses in certain ways.

A great deal of entrepreneurs' volunteering to benefit the local community emerges from business-specific knowledge of community needs; in other words, the nature of their businesses leads the owners to know about needs in certain corners of community life.[4] In helping to meet these needs, business owners donate not only their time but also, significantly, their expertise, gained from business experience in that place. Examples of this form of community involvement are legion; in their number and diversity these forms of volunteering suggest the many fascinating ways that business owners' volunteer activities are woven into the fabric of civic life. A partial list of these activities for Worcester appears in Table 10.1 and for Colorado Springs in Table 10.2.

In each of the cases listed in Tables 10.1 and 10.2 notice how running a business of a particular type or in a particular location leads the entrepreneur to understand a certain place-based need and then to take action to meet that need. These are specific instances of how entrepreneurs experience urban politics. So, for example, a man whose restaurant is located on a pond in Worcester works with a non-profit group to preserve the watershed of that pond; a woman whose retail store is located in a decaying downtown area directs a group working to restore that downtown. A woman who owns a kennel brings teens with Down Syndrome to her kennel to train

Table 10.1 Volunteer activities linked specifically to business: Worcester entrepreneurs

- Woman owner of a dance studio: gives dancing lessons to kids who can't afford them
- Male owner of shop that sells balloons and costumes, and provides clowns and entertainment: gives magic shows for Why Me (kids with cancer)
- Woman-owned kennel: her kennel is a training site for teens with Down Syndrome; she helps train these teenagers for working with animals
- Male owner of a construction firm that builds single-family homes: on the board of the Central Massachusetts Housing Alliance, an organization that works to meet the needs of the homeless and near homeless
- A couple that owns a site excavation firm: donates services to Habitat for Humanity and to the local town (e.g., to help build a playground)
- Woman-owned yoga studio and massage therapy: offers stress reduction workshops and free services for staff at the Rape Crisis Center and HIV/AIDS clients
- Woman owner of a funeral home: hospice volunteer
- Woman-owned sewing machine sales and service centre and provider of sewing lessons: organizes groups of customers to make 'comfort caps' for cancer patients, quilts for premature babies, and coats for homeless children
- Woman owner of a graphics, web design and publishing firm: started a free website open to any Worcester non-profits so these non-profits can operate their websites at no cost; is in the process of starting a non-profit that would put computer training centres in public housing projects and provide training for children and adults living there; organized the restoration of several historic public buildings in Worcester
- Woman owner of a human resources consulting firm: chair of the board of Children's Friend, a Central Massachusetts non-profit that serves the needs of children and their families
- Woman owner of a firm that provides services for the elderly: is a major fundraiser for Community Healthlink, a Central Massachusetts organization that provides health services for low-income people
- Woman owner of a retail children's clothing store located in the downtown area of an outlying town: directs a group working to restore the downtown
- Male owner of a restaurant on a pond within the City limits: volunteers with the non-profit group working to protect the pond's watershed; provides group with food and drink for their meetings
- Male owner of a commercial and residential real estate company: works with the Blackstone Valley Heritage Committee, which has brought development to the region
- Woman-owned commercial real estate firm: active in local non-profit groups promoting brownfield remediation and clean up of rivers

them in working with animals; a couple who run a site excavation firm donate excavating services to Habitat for Humanity (a national organization of volunteers who construct homes for low-income people) and to their local town for public space projects such as a playground. A woman owner of a web design firm launched a free website that is open to any Worcester-based non-profit; and a woman who owns a retail store selling horseback riding gear and Western clothing gives therapeutic riding lessons to disabled children. You can trace each of the volunteer activities noted in the examples in Tables 10.1 and 10.2 to the business owner's knowledge of a specific need, knowledge that, in turn, can be traced to the location of the business and/or the type of business.

Table 10.2 Volunteer activities linked specifically to business; Colorado Springs entrepreneurs

- Women owners of a pet grooming business: provide free grooming for animals at the Humane Society
- Woman owner of a horse equipment and western supply store: gives therapeutic riding lessons to disabled children
- A couple who own and operate a recreational volleyball facility: launched a competitive volleyball program for teenagers
- Male co-owner of an assisted living facility for the elderly: serves on the board of a local organization to provide a wide array of services to the elderly
- Woman owner of a high-end art gallery: with her husband, maintains the chapel in a historic church
- Male dentist: runs a dental program for an organization serving the disabled
- Woman practitioner of alternative Chinese medicine: volunteers her services at a hospice
- Male family practice lawyer: started a local program to advocate for children's rights
- Woman practitioner of alternative medicine and massage therapy: uses her skills (massage in particular) with children in Head Start and people in hospice care
- Woman bail bond poster: is a court appointed advocate for children
- Woman whose business runs corporate training programs: works with low-income minority children through the YMCA
- Man who runs a company (with his wife) that provides bus tours for the elderly: teaches computer classes at the senior center
- Man who co-owns a retail toy store with his wife: works with at-risk youth through a police department program
- Woman whose firm provides benefit packages to businesses: helps to refurbish houses for the indigent
- Woman optician: donates eyeglasses to the homeless shelter and other organizations serving low-income people
- Male owner of a printing firm: donates posters for various community events and fund drives

These activities take place in a wide array of perhaps unexpected locations, highlighting a theme of this volume that urban politics may be evident in unexpected spaces. They range from the entrepreneur's home space (e.g. horseback riding ring) or place of business (e.g. dance studio, kennel, yoga studio, sewing machine store, print shop) to local offices of non-profit groups, urban public spaces, the courtroom, hospices, churches, senior centres, rape crisis centres and other social service sites, as well as the virtual space of the internet. These urban politics take place in and through personal relationships, which construct a web of connections between and among individuals and groups located throughout the urban area. Take as one example the Worcester woman who owns a commercial real-estate firm and works with non-profit groups to clean up local rivers and brownfield sites: her volunteer activities link her not only to her co-volunteers in these efforts (who, in turn, live and work in many different locations around the urban area) but also to the sites targeted for clean-up.

It may be easy to dismiss these activities as being too trivial and too small in scale to count as a form of urban politics. But if political actions are indeed aimed at empowering the marginalized, challenging the status quo, or changing social or cultural practices, then these actions, individually and collectively, surely must count as a form of politics. In most of these examples the recipients are children, the elderly, or low-income people – all groups that include many who have not had a voice in the conduct of their everyday lives. Many of these examples describe actions aimed at building the human capital of individuals in these marginalized groups: working to increase the skills of low-income or at-risk children; working to improve the health and well-being of children, the elderly, or low-income families via education, health care, or housing. Many of the actions described in Tables 10.1 and 10.2 are aimed specifically at reducing inequality, which challenges the status quo. Increasing human capital and reducing inequality are likely to increase voice among those who have lacked voice in community affairs. Actions aimed at restoring environmental quality entail changes in social and cultural practices.

In addition to these forms of political action, some women entrepreneurs challenged the status quo in a number of other ways. One was caring for young and old family members – their own or those of employees – in their place of business, outside the home; in doing so, these business owners made visible the invisible and changed expectations about what is possible. Other women entrepreneurs were challenging convention simply by launching a business in a field typically dominated by men; Hanson and Blake (2005) describe many such instances, in which women changed opportunity structures for women by running businesses in, for example, construction, auto repair, engineering, or trucking.

Finally, it was clear from the interviews that many entrepreneurs, regardless of the type of business they ran, were committed to creating opportunities for others to start successful businesses. More than two-thirds of the entrepreneurs in this study said they mentored other prospective business owners, and this mentoring had a distinct gender dimension, with women being far more likely than men to mentor other women, and men emphasizing the mentoring of other men. Figure 10.1 suggests that this mentoring process has an important geographic dimension.

Bygrave and Minniti (2000) have theorized that, because entrepreneurship is a social process, the number of existing entrepreneurs in a person's vicinity affects the probability of that person becoming an entrepreneur. That is, proximity increases the chance of interacting with, being influenced by, and learning from, someone who already runs a business. But the characteristics of those entrepreneurs (such as gender) also affect the likelihood of social interaction (McPherson et al., 2001). In short, not only are role models important; the gender and geographic density of those role models are key to the mentoring process. As Figure 10.1 illustrates for the case of Worcester, the density of women-owned firms varies substantially at the census tract level, such that women in the highest-density tracts are more likely to encounter other women business owners and potential mentors than are women living in the northern or western part of the region. The suburbs east of the City with high densities of women's businesses are notable for their well-organized networks of, and

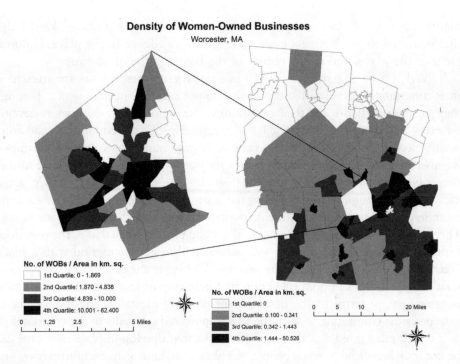

Figure 10.1 Number of woman-owned businesses/area per census tract, Worcester City (left) and metro area (right)

Source: Purchased list of all woman-owned businesses in the Worcester metro area, 1998.

acceptance of, women entrepreneurs (Hanson and Blake, 2005), which doubtless contribute to the high densities observed there. The map and its implications for mentoring are another way of thinking about the gendered social relations and spaces of political activity through which entrepreneurs affect the contours of opportunities for people in a place.

Discussion: Spaces of urban politics

This brief look at how entrepreneurs engage with the state and with other spaces of urban politics suggests that a focus solely on interactions with, and impacts on, formal government agencies and institutions misses significant ways in which business owners act politically to contribute to change. As noted at the outset, the conventional view holds that entrepreneurs transform places primarily through developing innovations, creating jobs, and paying taxes; and it is true that entrepreneurs do create important societal change in these ways. But they also transform places through volunteer activities that develop out of their knowledge of place and of people and their needs in that place – knowledge that is linked to the business owner's length of residence in the place, commitment to the place, and to the nature of the business she or he runs.

Rooted in place in part by owning a business there, entrepreneurs are invested in their communities in ways that those who do not own businesses are not. Through their business activities and their community connections, business owners develop business- and place-based knowledge of certain aspects of a community and community needs. Through their involvement in the community via volunteer activities – whether in formal organizations or not – they are forging connections and further deepening their embeddedness in place (Figure 10.2). Through many of these activities they build human capital among the least-empowered members of their communities, and in doing so, they create positive change and challenge the status quo. Through these activities, entrepreneurs also build place capital, that is, the resources/benefits that accrue to people or organizations through association with a place. Figure 10.2 summarizes this discussion and highlights the central role of personal relationships as sites of political action that help create place capital. Perhaps most notable here, in terms of the development of place capital, is the importance of entrepreneurs connecting with to the less-empowered segments of the community.

It is well recognized that the social networks that develop in place over time can be not only beneficial to some people, but also, simultaneously, exclusionary of, and detrimental to, less-powerful people and groups (Bourdieu, 1985). Feminist thinking

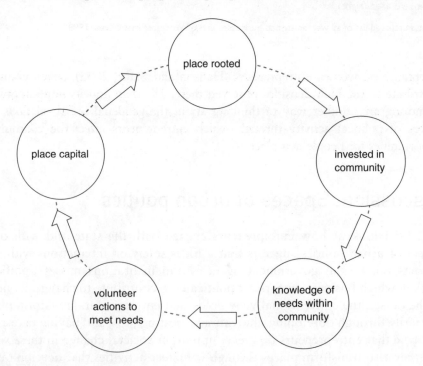

Figure 10.2 A virtuous circle? How place rootedness can lead to change

about politics directs attention to the central role of power and to those actions aimed at altering power relations. Through the diverse, small-scale volunteer activities such as those outlined in Tables 10.1 and 10.2, many entrepreneurs are changing power relations not in the voting booth or through a seat on the city council, but through personal relationships that connect them with people who have lacked political voice.

Acknowledgements

Suggestions from the editors on an earlier draft substantially improved this chapter. Thanks to Cynthia Adom and to Bill Hegman for help with Figure 10.1 and to Katherine Rugg for assistance in creating Figure 10.2.

Notes

1 Kofman (2005: 527) also notes that a well-known traditional political geographer, Norman Pounds, pointed out in 1972 that politics pervades people's everyday actions.
2 Note that woman-owned businesses were over-sampled; about one-third of privately owned businesses in the US are woman owned.
3 Registration as a WOB is likely to matter for firms that bid for government contracts, where women-owned and minority-owned businesses are given special consideration.
4 In a few cases, the volunteer work came first, and the business grew out of that work. One example is a woman who had volunteered at the Salvation Army's store, selling second-hand clothing and furniture; in part through this volunteering, she recognized the need among the City's low-income population for inexpensive second-hand furniture and household goods and later opened a store selling these items.

References

Aronson, Robert (1991) *Self-employment: A labor market perspective*. ILR Press: Ithaca, NY.
Bourdieu, P. (1985) 'The forms of capital', in Richardson, J. (ed.) *Handbook for Theory and Research for the Sociology of Education*. Greenwood Press: New York. pp. 241–258.
Brady, H. (2002) Personal communication.
Bygrave, W. and Minniti, M. (2000) 'The social dynamics of entrepreneurship', *Entrepreneurship Theory and Practice*, 24: 25–36.
Clark, William and Dieleman, Frans (eds) (1996) *Households and Housing*. Rutgers University, Center for Urban Policy Research: New Brunswick, NJ.

Cope, M. (2004) 'Placing gendered political acts', in Staeheli, L., Kofman, E. and Peake, L. (eds) *Mapping Women, Making Politics*. Routledge: New York. pp. 71–86.

Hanson, S. (2003) 'Geographical and feminist perspectives on entrepreneurship', *Geographische Zeitschrift*, 91: 1–23.

Hanson, S. and Blake, M. (2005) 'Changing the geography of entrepreneurship', in Nelson, L. and Seager, J. (eds) *Companion to Feminist Geography*. Blackwell: Oxford. pp. 179–193.

Kodras, J. (1999) 'Geographies of power in political geography', *Political Geography*, 18(1): 75–79.

Kofman, E. (2005) 'Feminist political geographies', in Nelson, L. and Seager, J. (eds) *A Companion to Feminist Geography*. Blackwell: Oxford. pp. 519–533.

Krugman, P. (2012) 'Moochers against welfare', *New York Times*, 16 February.

Malecki, E. J. (1994) 'Entrepreneurship in regional and local development', *International Regional Science Review*, 16(1–2): 119–153.

Malecki, E. J. (1997) 'Entrepreneurs, networks, and economic development: A review of recient research', in Katz, J. and Brockhaus, R. (eds) *Advances in Entrepreneurship, Firm Emergence and Growth*. JAI Press Inc: Greenwich, CT. pp. 57–118.

Martin, D., Hanson, S. and Fontaine, D. (2007) What counts as activism? The role of individuals in creating change, *Women's Studies Quarterly*, 35(3 & 4): 78–94.

Martin, P. (2004) 'Contextualizing feminist political theory', in Staeheli, L., Kofman, E. and Peake, L. (eds) *Mapping Women, Making Politics*. Routledge: New York. pp. 31–48.

McPherson, M., Smith-Lovin, L. and Cook, J. (2001) 'Birds of a feather: homophily in social networks', *Annual Review of Sociology*, 27, 415–444.

Moore, D. P. (1990) 'An examination of present research on the female entrepreneur – suggested research strategies for the 1990s', *Journal of Business Ethics*, 9(4 & 5): 275–281.

Peake, L. (1999) 'Politics', in McDowell, L. and Sharpe, J. (eds) *Glossary of Feminist Geography*. Arnold: London.

Rancière, J. (1999) *Disagreements*, trans. Julie Rose. University of Minnesota Press: Minneapolis, MN.

Scott, J. W. (1992) 'Experience', in Butler, J. and Scott, J. (eds) *Feminists Theorize the Political*. Routledge: New York. pp. 22–40.

Staeheli, L., Kofman, E. and Peake, L. (2004) 'Mapping gender, making politics: toward feminist political geographies' in Staeheli, L., Kofman, E. and Peake, L. (eds) *Mapping Women, Making Politics*. Routledge: New York. pp. 1–14.

Tilly, Charles (1990) *Coercion, Capital and European States, AD 990–1992*. Blackwell: Oxford.

West, C. and Zimmerman, D. (1991) 'Doing gender', in Lorber, J. and Farrell, S. (eds) *The Social Construction of Gender*. Sage: London. pp. 13–37.

11

IS CLASS RELEVANT TO URBAN POLITICS?

Mark Davidson

Introduction

This chapter considers the salience of social class for the study of urban politics. It does so in a paradoxical context. On the one hand, today we find an absence of class politics; old ideological battles about class are seen as archaic (Clark and Lipset, 1991; Giddens, 1999). Yet, on the other hand, we find the word 'class' saturating urban political discourse, including Richard Florida's (2004) (in)famous creative class. Perhaps the simple conclusion to be drawn from this paradox is that we are actually talking about different things. The former referencing class in a Marxian antagonistic relationship and the latter viewing class as something much more banal and, even, something to be embraced (at least in its creative form). It is the contention of this chapter that we are not dealing with such a simple shift. Rather, this paradox is explained by a post-industrial urban transformation that has complicated the ways in which we must understand social class.

Since the 1990s there has been a general acknowledgement that urban politics have changed. Before this point in time, growth regime theory (e.g. Stone and Sanders, 1987) was commonly used to explain urban politics (see Ward in this collection). This theory saw urban politics as being dominated by various business interests that cooperated with the state to deliver economic growth and related public goods. Local business leaders were therefore viewed as being central in constructing urban political institutions and development agendas since they held the resources required by city government:

> regime theory recognizes that any group is unlikely to be able to exercise comprehensive control in a complex world. Regime analysts, however, do not regard governments as likely to respond to groups on the basis of their electoral power or the intensity of their preferences as some pluralists do. Rather, governments are driven to cooperate with those who hold resources essential to achieving a range of policy goals. (Stoker, 1995: 59)

As a result, the conflicting interests of business owners and workers were considered a significant feature of urban political conflicts (e.g. Castells, 1977; Harvey, 1989; Saunders, 1981). With the advent of globalization and consequent significant industrial restructuring in many cities across the Global North (see Logan and Molotch, 1987) this theoretical explanation has become less popular.

Globalization and related changes in economic geographies have brought about new urban political landscapes. An important element of this change has been the different role of both business and workers in urban politics. As Clark and Harvey (2010) have argued '[C]hanges related to globalization have [...] fostered the *declining explanatory power of race and class in urban politics*' (424, emphasis in original). This globalized context is seen to have made city governments much more concerned with the need to compete with other cities in order to ensure existing businesses stay put and attract new capital to generate growth (Harvey, 1989). Importantly, this was not a shift decided upon by city governments or city-based business interests themselves. Rather, this has been brought about by wider economic and political changes (ibid.). Put simply, cities have had no choice but to compete with one another. The impacts of this shift have included the reduced influence of local political groups on urban government decision making (Brenner and Theodore, 2002), the decline of labour representation in urban politics (Boddy and Fudge, 1984) and the generation of new coercive, disciplining influences on local government decision making (Harvey, 1989).

This is 'new urban politics'. This form of urban politics is defined by 'cities or communities competing for mobile capital' (Cox, 1995). Within this mode of urban politics, the issue of class conflict is often seen as becoming less and less a central feature:

> Class politics emerged with industrialization. Labor union and socialist parties who opposed the hierarchy of industrial management characterized this era. The globalization of production, the emergence of new political actors, the development of post-industrial economies, and other processes related to globalization, have altered the explanatory power of these variables. (Clark and Harvey, 2010: 426)

In the entrepreneurial, post-industrial city, class antagonisms are therefore viewed by some as becoming less and less relevant.

This chapter critically examines this key premise of 'new urban politics'. It draws upon two case studies to illustrate both the relevance of class to urban politics and, in tandem, highlights the necessity to rethink the way in which social class is related to urban politics. The first example draws upon recent climate change debates and related policy making in London. This example is used to illustrate how questions of territory and boundaries (Ward and McCann, 2011) are directly related to questions of social class. The second example again draws on London, but this time the city is used to show how the ways in which we previously understood class relations have become problematic. Together these examples are used to argue that social class remains a central component of urban politics, albeit in ways that differ significantly from the ways it was in the mid-twentieth century.

Post-Industrialism and Social Class

Post-industrialisation brought with it a new urban landscape of social class. For some this represented a new epoch of social relations that were becoming largely devoid of class conflict (see Bell, 1973). Without a working-class population fighting against an exploiting capitalist class, it seemed some harmonious form of liberal, democratic capitalist society might emerge (Fukuyama, 1992). Elements, if not full-blown expressions, of this imaginary have been evidence across the social sciences. Indeed, critical philosopher Slavoj Žižek (2006) sees this idea of a society devoid of fundamental social class antagonisms (i.e. we are now left with merely technocratic issues to resolve) as a baseline assumption for many political theorists. But does this apply to the study of urban politics? Have we also assumed that, with the declining significance of political debate divided along class lines, the antagonistic relations between social classes have disappeared from the cityscape?

We can approach these questions by asking what the loss of the industrial city in many parts of the Global North has meant for urban politics. Numerous studies (for a summary see Imrie and Raco, 2003; Ross and Levine, 2011) have shown that politics have become less defined by class divisions (i.e. political parties are less divided along class lines) and that people have come to associate themselves with other identity categories (i.e. people are less likely to hold a strong class identity). This shift was captured by the social theorist Andre Gorz where he attempted to narrate the sociological impact of early post-industrial changes:

> In contrast to the proletariat in Marx's theory, the neo-proletariat does not define itself by reference to 'its' work and cannot be defined in terms of its position within the social process of production. The

question of who does or does not belong to the class of productive workers – how to categorise a kinesitherapist, a tourist guide, an airline employee, a systems analyst, a technician in a biological laboratory or a telecommunications engineer has no meaning or importance when set against a growing and more or less numerically dominant mass of people moving from one "job" to another.' (Gorz, 2001 [1980]: 70)

Gorz's main point here is that changes in occupational structures have meant it has become difficult to identify a particular working-class group (i.e. proletariat) and, consequently, distinct class groupings and conflicts. But Gorz is not saying class relations have disappeared. He is arguing at these relations still exist, but in different forms. Gorz sees the post-industrial city as bound up with class division and proletarian exploitation, but in distinct ways from industrial cities.

This interpretation of changes in social class contrasts to other more sanguine interpretations of post-industrialism (Bell, 1973; Fukuyama, 1992). Gorz rejects the idea that post-industrial societies are becoming devoid of class antagonisms whilst at the same time attempting to make sense of new economic and occupational structures. He certainly thought traditional working-class politics – such as those organized around some industrial worker/masculine identity – were a thing of the past; much to the chagrin of many socialists who held on tightly to this model of politics. So, for Gorz, class politics was still in evidence but just now infused quite differently across the new occupations that collectively defined the socio-economic structure of the post-industrial city.

Gorz's (1980) interpretation of post-industrial class relations connects with more recent attempts in political philosophy to theorize class in the absence of distinct class divisions. This has involved a return to some foundational theoretical premises. Marx made it perfectly clear that capitalism, in its most simple abstraction, involved an antagonistic relationship between capital and labour. Through his use of the labour theory of value Marx argued that capital re-created itself (and capitalists became rich) by stealing labour time from labourers. This interpretation of the labour theory of value, a theory which Marx borrowed from mainstream economic theory, would eventually push classical economists to abandon the theory. Although the labour theory of value and its associated reading of the relationship between labour and capital are theoretical abstractions, the industrial city would come close to reflecting them. Direct conflicts between large groups of workers (i.e. labour) and business owners (i.e. capital) defined many urban political landscapes in the early twentieth century. However, a combination of the state management of economy – and the related provisioning of social welfare services – and later declines in heavy industrial manufacturing in places such as the USA and UK removed much of this political landscape. With the advent of Thatcherism/Reaganism in the 1980s and its subsequent reformulation under Tony Blair and Bill Clinton, it seemed that class divisions were irrelevant to governments who thought that little conflict existed between the interests of business and citizens (see Harvey, 2005).

This gap between appearance (i.e. how class relations look) and what exists (i.e. the theoretical understanding of a fundamental antagonism between capital and labour) has been of much interest to philosopher Slavoj Žižek (2001). He has argued that much is at stake with regards to how we understand this relationship. For elites, Žižek argues, it is in their interests to produce a social understanding – an ideology – that does not fully recognize or acknowledge what exists: 'every hegemonic universality has to incorporate *at least two* particular contents: the 'authentic' popular content and its 'distortion' by the relations of domination and exploitation' (184). What Žižek is saying here is that capitalism has this antagonistic relation between capital and labour within it, but that this is rarely or never fully symbolized. Our view of it is always mediated, usually in a way that downplays its presence or even existence. If we accept the presence of an antagonistic relation between capital and labour we will therefore always have a politics that, in some way, reflects this. Furthermore, there will always be a requirement to obscure this relation for those who benefit from these social relations.

However, Žižek (2001) claims that the obscuring of class antagonisms is not just carried out by elite business interests. He has claimed that the class antagonism is also obscured by many other actors. These include the middle classes, those people who have been seen to now dominate the socio-economic structure of the post-industrial city (Butler et al., 2008; Florida, 2004). Žižek (2001) argues that the middle classes, or rather their particular self-identification as middle class, play a key role in the obscuring and denial of class antagonisms: 'the only class, which, in its "subjective" self-deception, explicitly conceives of and presents itself as a class is the notorious 'middle-class' which is precisely the "non-class"…' (186). Whilst the (non-)presence of certain social classes in the post-industrial context has led to many to diagnose a decline in its relevance to urban politics (Clark and Harvey, 2010), here Žižek points to the paradoxical prominence of the middle class as a sign of class politics. He is arguing that this group is, in a purely theoretical sense, incompatible with the ideas of capital and labour. When Žižek claims the middle classes present themselves as the social whole (i.e. the post-industrial utopia of an almost entirely middle class society), he sees this as 'the denial of antagonism' (ibid. 187). Put differently, Žižek is saying that it is impossible to have a purely middle-class society that is capitalist, since capitalism requires people who will (a) be the owners of capital and (b) be the labourers who are exploited. To present the idea that an entire capitalist society might become middle class denies this simple theoretical conclusion.

This line of thought is pushed further in Žižek's (2001) analysis when he claims that we must understand class politics as bound up with processes of concealment. He argues:

> Leftists usually bemoan the fact that the line of division in the class struggle is a rule blurred, displaced, falsified ... However, this constant displacement and 'falsification' of the line of (class) division *is* the 'class struggle': a class society in which the ideological perception of the class division was pure and direct would be a harmonious structure with no struggle ... (ibid., 187)

This argument forces us to think carefully about social class in urban politics. It challenges us to examine whether class struggle has purely disappeared/declined (i.e. the abstract antagonism between capital and labour is resolved) or ask whether it is merely 'blurred, displaced, falsified' (ibid., 187). Furthermore, it pushes us to be concerned with class politics in the ideological domain; towards those various attempts to symbolize and explain class relations in ways that avoid the antagonism becoming visible.

We can therefore see Gorz and Žižek in agreement with respect to how class relations are not related to any particular type of party politics and/or occupational structure. What Gorz (1980) argues is that the changing presence of the stereotypical 'working classes' (i.e. the workers of the industrial city) cannot be seen as evidence of the decline of class politics itself. His book *Farewell to the Working Class* is not an abandonment of class politics, rather it is an attempt to articulate what the class antagonism means in post-industrial societies that do not have large traditional working class populations and associated political organizations. As class-based politics (i.e. voting patterns, political parties, labour movements) have declined in many countries, some commentators have been quick to read this as a declining importance of class analysis per se. Yet this change should not be read as a de facto loss of antagonistic class relations. We therefore do not need a theory of urban politics that looks at social class as peripheral, but rather one that seeks to understand how class relations are now constituted, enacted and displaced.

In the following sections two examples of the mutated presence of social class in urban politics are illustrated. The first section outlines the ways in which current understandings of socio-spatial relations are important with respect to theorizing urban politics. This involves a questioning of the often foundational idea of cities as territorially bound. The second section turns to the question of class identities in the post-industrial city. The central concern here is the problem of reading a decline in working-class occupations/identities as a decline of the class antagonism.

Out of bounds urban politics

In Doreen Massey's *World City* (2007), a book that examines post-industrial London, she calls for the city to develop 'a politics of place beyond place' (188), arguing that cities need an inverted form of localism. She argues '... "place" would seem to have real, and maybe ironically, in this age of globalisation, even increasing potential as a locus of political responsibility and an arena for political engagement' (208). Massey's point is that a city's internal politics must become extroverted and, in doing so, connect internal political debates (e.g. local struggles of housing provision, transit planning) to more global concerns (e.g. poverty, climate change, uneven development). Here cities are '... meeting-places of multiple trajectories' (207) and therefore urban politics should extend through these trajectories. Whilst Massey recognizes the complexity of any such analytical move, she argues this would 'highlight the structural connections between inequality at a global level and the inequality within the city' (207). Urban politics therefore becomes about role in a global set of processes. The city is a staging post of incredibly complex and interweaving social relations.

This represents a significant shift in urban political theory. Whilst 'urban politics' has never been a coherent concept, it has been continually reformulated around ideas of scale and spatial form. For example, John (2009: 17) recently offered the following starting point: 'At its most straightforward, urban politics is about authoritative decision-making at a smaller scale than national units ... the focus of interest is at the sub-national level with particular reference to the political actors and institutions operating there.' In this case, the territory of the city serves as the definition: urban politics are those politics that take place within the city.[1] This premise of a smaller scale unit of politics operating within other sets of hierarchical/nested units has traditionally been a dominant one in the urban politics literature (Castells, 1977; Harvey, 1989; Saunders, 1981). Yet this schema is problematic in Massey's framework since it reduces urban politics to a certain scale that does not necessarily connect to, or constitute, global processes (also see Allen and Cochrane, 2010; Ward and McCann, 2011).

An example of how Massey's (2007) framework has a different analytical emphasis can be demonstrated by thinking about London's urban regime, that set of business and political interests that are pivotal to the city's successes and failures. Viewed with an extroverted sense of place, this set of actors must be seen as operating both within London *and* beyond the city's boundaries. In London's case this would see urban politics played out with reference to how powerful financial interests within the City of London (see Figure 11.1) play a central role in producing and coordinating flows of capital across the globe. Where these flows go and what they generate would therefore be concerns for Londoners. To restrict urban politics to local electoral boundaries denies urban politics the extrovert dimension Massey seeks to elevate.

Figure 11.1 Bank Junction, City of London: a key site of global financial networks (source: author)

The type of urban politics Massey (2007) theorizes is, to some degree, already present in urban policy debates. This can be illustrated by looking at policy debates around climate change. In the Greater London Authority's attempt to develop a sustainability agenda, London's mayoral government has had to think about local urban political changes in the context of a global environmental issue. In London's recent draft environmental strategy policy, *Delivering London's energy future* (GLA, 2010), the issue of climate change has required that urban political concerns be extroverted, even if the complete set of political relations that might come with such a perspective have not been examined (see Box 11.1).

> **BOX 11.1 COMMENTARY ON LONDON'S ENERGY POLICY**
>
> London's energy policy document *Delivering London's Energy Future* (2010) deals with the potential impacts of climatic change by seeing the change as an economic opportunity. The Mayor's foreword begins by stating:

> London is at the cusp of an exciting energy revolution. A potent combination of rising concerns over energy security and long-term increases in fossil fuel prices has led to a growing awareness that our traditional energy resources are finite. Meanwhile city living, especially in an expanding metropolis like London, leads to pollution that poses a threat to our health and quality of life. In addition, tackling climate change by reducing greenhouse gas emissions has become a major global priority, which requires urgent action. (5)
>
> The Mayor goes onto reframe this environmental problem as an economic necessity:
>
> > It is vital that our growing city develops and grows in a way that exemplifies greener living ... London is well positioned to seize the opportunities coming from this nascent low carbon age, to be one of the world's leading low carbon capitals and the leading carbon finance centre. But we cannot be complacent – other cities and countries are competing for this prize. (6)
>
> Whereas the urban politics of climate change might have revolved around the responsibility of cities in the Global North to reduce emissions and provide reparations for those affected by anthropogenic climate change, in London's energy policy document the issue is transformed into another entrepreneurial urban policy agenda.

London's climate change action plan (GLA, 2010) centres on perceiving climate change adaptation as an opportunity to grow the city's economy and maintain its world-city status. However, there is a persistent requirement within the document to recognize the global dimension of climate change-related greenhouse gas (GHG) emissions: 'Further action is therefore required, and although London's relative contribution to global GHGs is small, as a world city, it has an important leadership role to play in reducing emissions and moving to new models of energy generation and consumption' (ibid., 9). London's per capita emissions certainly rank lower than most other cities in the UK (Bicknell et al., 2009) and are similar to cities such as New York. However, an acknowledgement of the unsustainable level of these GHGs – 44.3m tons of CO_2 in 2006; some 8 per cent of the UK total and 6.18 tons per person (ibid.) – also invokes two extroverted political dimensions: (a) the disparity in responsibility for carbon emissions at the global level and (b) the relations that London's command and control functions (i.e. multi-national corporations and financial services) have with respect to the organization/facilitation of those industries that are overwhelmingly responsible for climate change.

An extroverted urban politics would embrace this geographically complex set of concerns by drawing attention away from local adaptation and towards the issues of global responsibility and global cooperation. It is perhaps therefore unsurprising then that Mayor Boris Johnson has downplayed issues such as London's disproportionate responsibility for climate change. Not only would this extroverted politics complicate the city's political landscape, it would also potentially politicize London's past and current unsustainable per capita levels of greenhouse gas emissions.

Of course, energy policies are just one part of a city like London's policy concerns. Yet just this one example demonstrates the different debates that might need to be had within an extrovert politics of place. London's external relations are innumerable. There are few places on the planet that are not, in some way, connected to activities going on within the city. London's financial services industry tends daily to the requirements of global capitalism, organizing its flows, participants, geographies, etc. These economic activities construct a large set of the city's extrovert relations. We might therefore ask how the politics of these external relations could be reflected in urban politics. If a politics of place must extend beyond the city's boundaries, what is it about London's financial relations that need to become articulated? Just as with energy policy and climate change, we might ask about the responsibility the city has for mortgage securities in the USA, for national debt in Greece, for debt burden in sub-Saharan Africa. It is precisely in these sets of concerns that we might find some of the urban politics of social class in an age of globalization.

Class identities in the post-industrial city

If we accept the idea of an extroverted urban politics, it is clear that our political consciousness has to change. We would be as concerned about the relations we have with people beyond the city limits as those relations we have within the city. Yet we cannot forget about the society forged between people living within a particular space. There is little doubt that the city itself represents a community, a grouping of people bound together through concerns such as infrastructure, transit, culture and environment.

Since the advent of post-industrialism (Bell, 1973) there has been a great deal of concern about the class constitution of the city. Debates have raged over whether the post-industrial city has progressively become more middle class (Hamnett, 1994) or whether it is more and more divided between rich and poor (Sassen, 1991). The implications of each reading are significant. This is perhaps best exemplified by the contrasting interpretations of gentrification. For some who see the post-industrial city as an increasingly middle-class community, gentrification represents the replacement of a now historic social formation (Butler et al., 2008; Haase et al., 2005). Neighbourhood transitions from working class to middle class are viewed as a replacement and re-population process. A declining presence of the traditional working classes (e.g. those registered in the census as skilled labourers) is read as a more general decline in the internal presence of antagonistic class relations. If we therefore remain solely concerned with urban politics as something contained within cities, this interpretation would see politics as less and less wrapped up with class since socio-economic status becomes more homogeneous.

There are two main problems with this interpretation. The first relates to the myopic focus on the city as a setting for politics which leads to a 'literal and figurative effacing of the proletariat' (Wacquant, 2008: 199) by overlooking the broader social relations cities are bound up in (see above). The second is concerned with the way we actually understand the class identity of people living in post-industrial cities. Whilst there has been a shift within the class composition of many cities, one that has often been understood as involving a decline of working-class presence and, consequently, a decline in class antagonisms, we should proceed with caution using this interpretation (see Watt, 2008). This caution is both methodologically and theoretically motivated.

In terms of theory, we can return to our discussion of social class earlier. For critical theorists, class relations are seen to stem from the antagonistic relationship between capital and labour (Žižek, 2001). As a result the narration of the middle-class city is a highly problematic assertion because it obscures that fact that the capitalist city is bound up in an antagonistic social relation. As sociologists have argued, the idea of being middle class is actually constructed from the counterpoising of working labourers and capitalists:

> The definition of 'middle class' is vague but evocative ... developed as a negative term [...]. By calling yourself middle class you distinguished yourself from those above you [...] and those below you ... But this does not indicate that different people within the middle classes actually have anything in common other than that they are not upper or lower class. (Savage et al., 1992: xi)

The point the authors above are making is that the notion of 'middle class' comes from the idea of a group of people being positioned between lower and higher classes. They identify 'middle class' as a vague term since it relies on there being something else either side. As such, if we identify a city such as London as increasingly middle class we might ask – in the absence of a non-capitalist economy – where is the class relation? On one hand, this might involve adopting a new geographical perspective on urban politics, something more akin to Massey's (2007) extrovert sense of place. On the other hand, we might revisit the issue of how we actually measure the class composition of the post-industrial city. This would be a methodological concern.

If we examine the way social class is measured in the UK census, we can see how difficult it is to clearly define someone's class status. In recent UK censuses, social class measures have been revised twice, the latest two versions being called SEG (socio-economic group) and SEC (socio-economic classifications) schemes (see Table 11.1). The conceptual frameworks used to develop these schemes draw heavily on John Goldthorpe's (2007) attempt to distinguish between different locations in the labour market. In this pragmatic attempt to divide up the labour market, Goldthorpe identifies a working class (i.e. skilled labourers) and a middle class (i.e. managers and professionals). As a consequence

Table 11.1 Operational categories of the National Statistics Socio-economic Classification (NS-SEC) linked to socio-economic groups

Socio-economic group		NS-SEC operational categories
1	Employers and managers in central and local government, industry, commerce, etc. – large establishments	
1.1	Employers in industry, commerce, etc. – large establishments	1
1.2	Managers in central and local government, industry, commerce, etc. – large establishments	2
2	Employers and managers in industry, commerce, etc. – small establishments	
2.1	Employers in industry, commerce, etc. – small establishments	8.1
2.2	Managers in industry, commerce, etc. – small establishments	5
3	Professional workers – self-employed	3.3
4	Professional workers – employees	3.1
5	Intermediate non-manual workers	
5.1	Ancillary workers and artists	3.2, 3.4, 4.1, 4.3, 7.3
5.2	Foremen and supervisors non-manual	6
6	Junior non-manual workers	4.2, 7.1, 7.2, 12.1, 12.6
7	Personal service workers	12.7, 13.1
8	Foremen and supervisors – manual	10
9	Skilled manual workers	7.4, 11.1, 12.3, 13.3
10	Semi-skilled manual workers	11.2, 12.2, 12.4, 13.2
11	Unskilled manual workers	13.4
12	Own account workers (other than professional)	4.4, 9.1
13	Farmers – employers and managers	8.2
14	Farmers – own account	9.2
15	Agricultural workers	12.5, 13.5
16	Members of armed forces	–
17	Inadequately described and not stated occupations	16

of using these categorizations in cities like London we have seen significant decreases in the numbers of those occupying working-class positions. This is quite reasonable. But what if the nature of working-class occupations has changed in these cities? What if our categories are not good indicators of class status?

This idea that we might have a new form of working-class population is nothing new. Indeed, we have seen a vast documenting of workers in industries such as food services and retail services (e.g. Ehrenreich, 2002) that, whilst being service-economy workers, are also subject to oppressive and exploitative working conditions. But oftentimes these occupations are not viewed as being working class in the sense that they do not fit into working-class categories in census measures. Our methodological problem is therefore concerned with who holds a working-class position in the absence of an economy that has clearly delineated class-related occupational divisions.

In London, the SEC groups that have seen the greatest growth since the 1980s are numbered 5.1 and 5.2. (see Butler et al., 2008). These can be considered lower-middle-class occupations and, consequently, these groups are often used to illustrate how the city has become more middle class and less working class (ibid.). However, if you examine the occupations in these classifications they are highly varied (see Rose et al., 2005 for a detailed discussion). They range from actors, assistant nurses, immigration officers, estate managers through to typists, debt collectors, cashiers, sales assistants and petrol pump forecourt attendants (see Rose and O'Reilly, 1998: 56–91). Clearly, some of these occupations are middle class. However, others, such as sales assistants, would seem not to be middle class. Even within the context of an archetypal post-industrial city such as London there is reason to question the extent to which class composition has changed.

We are therefore faced with two intertwined issues when considering the relationship between social class and urban politics. First, the continued fetishization of the city setting (i.e. it as a bounded political space) means that we need new conceptual approaches to think about how cities are bound up in class relations. As cities become globally interconnected in deeper and more extensive ways we will likely require an extroverted urban politics to capture the political dimensions of urban life. Second, although the post-industrial city is clearly different from its historical antecedents, we need to carefully consider how this transition has impacted social class composition. Whilst it has undoubtedly changed working-class occupations, political organizations and class consciousness, the extent to which this should be understood as a process of class homogenization is questionable. We therefore need to rethink the ways in which class antagonisms are both instilled and produced through the city *and* identify the class character of the post-industrial city in the context of a decline in traditional working-class occupations.

Conclusions

In the last two decades there has been a shift away from understanding urban politics as being organized around social class. To some extent this reflects the Anglo-American focus of the urban politics literature, since many cities in the world are not post-industrial and/or have a vibrant class politics (Roy, 2009). This chapter has attempted to demonstrate the necessity to remain concerned with the social class dimensions of urban politics but in a way that looks at questions of class through both the external relationality of cities and the transformed internal composition of occupational structures in the post-industrial city.

Yet we must not rest here. We must be careful with how we think about the post-industrial city itself. Identifying the post-industrial city as a next step in the city's historical development (e.g. Bell, 1973) can serve to depoliticize the city's class relations. Take the idea of new urban politics and the tendency to see class as unimportant in urban politics. It might certainly be the case that municipal elections and working-class politics have come to play a less prominent role in urban politics. But we must be careful with how we develop our understanding of this situation. If we remain focused on the city territory as the locale of urban politics, we eviscerate the class relations that are constitutive of it (Massey, 2007). This is not to say that a focus on internal politics is unimportant, but rather to stress that the constitutive processes of the city demand that our theoretical approach to urban politics be able to incorporate both our the internal and external considerations.

A (re)engagement with the social class dimensions of urban politics therefore requires substantial theoretical renewal. Here we can learn from recent debates in political philosophy. Žižek (2006) recently made the following point about how we develop knowledge: '*the bracketing itself produces the object*' (56; emphasis in original). His point was that the way in which we frame the object of inquiry – in this case urban politics – has important implications for the way in which the researcher approaches and perceives the object. In the case of urban politics we can take the notion of bracketing in the literal spatial sense by pointing to the continued reliance on the notion of bounded urban political space (e.g. John, 2009). Although not to claim this bracketing is redundant, it is to say it is highly problematic when used as the main entry point for thinking urban politics. And this is not purely for theoretical reasons. If we identify this bracketing of urban politics as obscuring the city's class constitution, the approach also must be seen as having a politics: '… it concerns what Marx called "real abstraction"; the abstraction from power and economic relations is inscribed into the very actuality of the democratic process' (ibid., 56). Here Žižek points to the politics of bracketing. The way we theorize urban politics engages in a procedure that can strip away many of the political aspects of the city. And if class politics is about the full representation of the capital/labour antagonism (Žižek, 2001, 2006), we must be much more aware about how certain approaches to urban politics lose sight of class relations in the capitalist city.

Note

1 Other examples include Davies and Imbroscio's (2009) defining of the themes of urban politics as 'who wields urban political power, the nature of urban governance, and how urban citizens both affect and are affected by these dynamics of power and governance' (5). In this outlining, the spatial framing of urban politics remains consistent and we switch focus across actors.

References

Allen, J. and Cochrane, A. (2010) 'Assemblages of State Power: Topological Shifts in the Organization of Government and Politics', *Antipode*, 42(5): 1071–1089.

Bell, Daniel (1973) *The Coming of Post-Industrial Society*. Basic Books: New York.

Bicknell, Jane, Dodman, David and Satterthwaite, David (eds) (2009) *Adapting Cities to Climate Change: Understanding and Addressing the Development Challenges*. Earthscan: London.

Boddy, Martin and Fudge, Colin (eds) (1984) *Local Socialism*. Macmillan: London.

Brenner, N. and Theodore, N. (2002) 'Cities and the Geographies of "Actually Existing Neoliberalism"', *Antipode*, 34(3): 349–379.

Butler, T., Hamnett, C. and Ramsden, M. (2008) 'Inward and Upward? Marking out social class change in London 1981–2001', *Urban Studies*, 45(2): 67–88.

Castells, M. (1977) *The Urban Question: A Marxist Approach*. MIT Press: Cambridge, MA.

Clark, T. and Harvey, R. (2010) 'Urban Politics', in Leicht, K. and Jenkins, C.J. (eds) *Handbooks of Politics: State and Society in Global Perspective*. Springer: London.

Clark, T. and Lipset, S. (1991) 'Are Social Classes Dying', *International Sociology*, 6(4): 397–410.

Cox, K. (1995) 'Globalisation, Competition and the Politics of Local Economic Development', *Urban Studies*, 32(2): 213–224.

Davies, J. and Imbroscio, D. (2009) 'Urban Politics in the Twenty-first Century', in Davies, J. and Imbroscio, D. (eds) *Theories of Urban Politics*, Second Edition. Sage: London. pp. 1–14.

Ehrenreich, Barbara (2002) *Nickel and Dimed: On (Not) Getting By in America*. Holt: New York.

Florida, Richard (2004) *Rise of the Creative Class*. Basic Books: New York.

Fukuyama, Francis (1992) *The End of History and the Last Man*. Free Press: New York.

Giddens, Anthony (1999) *The Third Way: The Renewal of Social Democracy*. Polity: London.

Goldthorpe, John (2007) *On Sociology Second Edition Volume Two: Illustration and Retrospect*. Stanford, CA: Stanford University Press.

Gorz, André (2001 [1980]) *Farewell to the Working Class*. Pluto Press: New York.

Greater London Authority (GLA) (2010) *Delivering London's Energy Future: The Mayor's Draft Climate Change Mitigation and Energy Strategy for consultation with the London Assembly and Functional Bodies*. Greater London Authority: London.

Haase A., Kabisch S. and Steinführer A. (2005) 'Reurbanisation of Inner-city Areas in European Cities' in Sagan, I. and Smith, D. (eds) *Society, Economy, Environment – Towards the Sustainable City*. Bogucki Wydawnictwo Naukowe: Gdansk. pp. 75–91.

Hamnett, C. (1994) 'Social Polarisation in Global Cities: Theory and Evidence', *Urban Studies*, 31(3): 401–424.

Harvey, D. (1989) 'From Managerialism to Entrepreneurialism: The Transformation of Urban Governance in Late Capitalism', *Geografiska Annaler*, 71(B): 3–17.

Harvey, David (2005) *A Brief History of Neoliberalism*. Oxford University Press: Oxford.

Imrie, R. and Raco, M. (eds) (2003) *Urban Renaissance? New Labour, Community and Urban Policy*. Policy Press: Bristol.

John, P. (2009) 'Why Study Urban Politics?', in Imbroscio, D. and Davies, J. (eds) *Theories of Urban Politics*, Second Edition. Sage: London.

Logan, J. and Molotch, H. (1987) *Urban Fortunes: The Political Economy of Place*. University of California Press: Berkeley, CA.

Massey, Doreen (2007) *World City*. Polity Press: London.

Robinson, Jennifer (2006) *Ordinary Cities: Between Modernity and Development*. Routledge: London.

Rose, David and O'Reilly, Karen (1998) *The ESRC Review of Government Social Classifications*. Office for National Statistics: London.

Rose David, Pevalin, David and O'Reilly, Karen (2005) *The NS-SEC: Origins, Development and Use*. Palgrave: Basingstoke.

Ross, Bernard and Levine, Myron (2011) *Urban Politics: Cities and Suburbs in a Global Age*, Eighth Edition. M.E. Sharpe: New York.

Roy, A. (2009) 'The 21st-century metropolis: new geographies of theory', *Regional Studies*, 43(6): 819–830.

Sassen, Saskia (1991) *The Global City: New York, London, Tokyo*. Princeton University Press: Princeton, NJ.

Saunders, Peter (1981) *Social Theory and the Urban Question*. Unwin Hyman: London.
Savage, Mike, Barlow, James, Dickens, Peter and Fielding, Tony (1992) *Property, Bureaucracy and Culture: Middle Class Formation in Contemporary Britain*. Routledge: London.
Stoker, G. (1995) 'Regime Theory and Urban Politics', in Judge, D., Stoker, G. and Wolman, H. (eds) *Theories of Urban Politics*. Sage: London. pp. 54–71.
Stone, Clarence and Sanders, Heywood (1987) *The Politics of Urban Development*. University Press of Kansas: Lawrence, KS.
Wacquant, L. (2008) 'Relocating gentrification: the working class, science and the state in recent urban research', *International Journal of Urban and Regional Research*, 32(1): 198–205.
Ward, K. and McCann, E. (2011) 'Cities Assembled: Space, Neoliberalization, (re) Territorialisation, and Comparison', in McCann, E. and Ward, K. (eds) *Mobile Urbanism: Cities and Policymaking in the Global Age*. University of Minnesota Press: Minneapolis, MN. pp. 167–186.
Watt, P. (2008) 'The Only Class in Town? Gentrification and The Middle-class Colonization of the City and the Urban Imagination', *International Journal of Urban and Regional Research*, 32(1): 206–211.
Žižek, Slavoj (2001) *The Ticklish Subject*. Verso: London.
Žižek, Slavoj (2006) *The Parallax View*. MIT Press: Cambridge.

12

THE URBAN IMAGINARY OF NATURE: CITIES IN AMERICAN ENVIRONMENTAL POLITICS[1]

Matthew Huber

Introduction

> We see nature through the geographical and historical experience of the urban. (Fitzsimmons, 1989: 108)

Though rarely acknowledged, contemporary politics of nature have emerged in an urbanized and urbanizing world. The environmental problematic has risen to the centre of political struggles over urban governance related to such concerns as climate change, urban forest cover, parks, and water systems. Much of this politics emerges out of a recognition that the urban socio-ecological metabolism is responsible for the majority of consumption of energy, resources and land, and the multitude of 'environmental' problems bound up with these patterns (Low 2000; Heynen et al., 2006).

Somewhat ironically, even as urban geographies produce large-scale ecological impacts, it is well recognized in the United States that the environmental movement has gained its political strength mainly from urban and suburban populations (Dowie, 1995; Rome, 2001). In his history of post-World War II US environmental politics, the prominent historian Samuel Hays asserts that 'environmental values are an integral part of an urbanized society ... grow[ing] as urbanization has grown' (Hays, 2000: 23–24). Hays's claim is based upon the fact that mainstream environmental political organizations, such as the Sierra Club, the Wilderness Society and the Audubon Society, draw their membership (and thus financial) base from urban and suburban areas in the United States (Bosso and

Table 12.1 The changing nature of American environmental politics

Text	Idea of nature	Politics
Aldo Leopold, *Sand County Almanac*	The biotic community wilderness	Anti-urban, Malthusian land ethic
Rachel Carson, *Silent Spring*	Nature as 'sink' Repository environment	National regulation of pollution
Al Gore, *An Inconvenient Truth*	Global biosphere wilderness	Anti-urban, Malthusian global regulation of pollution individualized consumption

Guber, 2003). Even more confusing is the fact that American environmental politics is also structured through a particular 'anti-urban' construction of nature and, in particular, wilderness (Cronon, 1995). This paradox between a specifically urbanized anti-urban environmental politics raises the question: what is particularly urban about mainstream environmental politics, and how has an *urban* imaginary of nature become politically and economically powerful in the construction of environmental politics both in and outside the city?

In this chapter I situate what Raymond Williams (1980: 67) famously called 'ideas of nature' within the processes of urbanization. Specifically, I utilize the concept of the 'imaginary' (Peet and Watts, 1996; Zukin et al., 1998; Gandy, 2006). The notion of 'imaginary' is simply a conceptual tool to understand how commonsensical socially constructed meanings become infused into fields of social power and material practice. Imaginaries are discursive regimes that endow meanings with the status of truth and/or moral righteousness (Foucault, 1978). As Zukin et al. (1998: 628) point out, 'although the imaginary derives from poststructuralist, psychoanalytic discussions of the unconscious, it is useful for demonstrating the social power exercised by cultural symbols on material forms'. The imaginary suggests a simultaneous engagement with the material and symbolic. Thus, I seek to position ideas of 'nature' as both imbricated in the historically specific processes of capitalist urbanization, and representing a potentially open, 'terrain' through which political challenges to those processes unfold.

For the purposes of this specific chapter, I build off notions of the environmental imaginary (Peet and Watts, 1996) and urban imaginary (Zukin et al., 1998), to articulate a particular 'urban imaginary of nature'[2] that locates the realm of the urban as the antithesis of 'nature' (e.g. Jacobs, 1961: 443–445). The starting point of this chapter is that the category of 'nature', itself, is socially constructed, or imagined (Castree and Braun, 1998; Demeritt, 2002). Specifically, I hope to understand how discursive and material aspects of an imagined nature are mobilized to construct, legitimate and indeed become emblematic of specific forms of environmental politics.

This chapter consequently centres around two basic questions. First, as Alexander Wilson (1992: 25) puts it from a historical standpoint, 'nature appreciation directly coincided with urbanization ...' Thus, how can ideas of nature be understood through the historical specificities of capitalist urbanization? This question will be informed through a re-reading of Simmel (1995 [1903]) and Wirth's (1938) classic texts on 'the urban question' and their assumptions about the relations between cities and nature. Second, how do urban imaginaries of nature get translated into politics? To examine this question I examine three quintessential texts that have informed the mainstream environmental movement in the United States – Aldo Leopold's (1949) *Sand County Almanac*, Rachel Carson's (1962) *Silent Spring*, and Al Gore's (2006) *An Inconvenient Truth*. I will argue that these three texts, in different ways, frame the city as a specific driver of ecological despoliation, and, thus, promote a politics of nature *against* cities. As such, the environmental political construction of nature can be situated within the longstanding traditions of anti-urban discourse in American history (e.g. Macek, 2006). I therefore argue that environmental politics *as a whole* are already *specifically urban* forms of politics emerging through the historical and material experience of urbanization itself.

The Denatured Metropolis and Mental Ways of Life

Georg Simmel (1995 [1903]) and Louis Wirth (1938) provide two classic takes on the 'urban question' (Castells, 1977; Brenner, 2000). Simmel (1995 [1903]: 45) claims the metropolis is overflowing with human and nonhuman stimuli through which the individual's only rational, adaptive response is that of the 'blasé outlook', or an attitude of indifference towards people and things and the distinctions between them. He asserts that the 'money economy' provides the only means of differentiating among 'things':

> To the extent that money, with its colorlessness and indifferent quality, can become a common denominator of all values it becomes the frightful leveler – it hollows out the core of things, their peculiarities, their specific values and their uniqueness and incomparability in a way which is beyond repair. (ibid., 36)

Simmel seems to imply that the monetized urban realm reduces the 'core' of things – and the relations through which production and exchange of things are mediated – into a single, objective quantitative determinant. Simmel also argues that individuals within the blasé, monetized urban realm are also subject to the 'objective' culture of the division of labour, situating them as only a one-sided 'cog' in the urban machine (44). This implies that the one-sidedness of city life may also yield a yearning imaginary for some deeper qualitative relationship with not only 'things' but human beings themselves.

Simmel goes on to argue that metropolitan social relations are reduced to a struggle amongst individuals in the absence of nature: 'The decisive fact here is that in the life of the city, struggle with nature for the means of life is transformed into a conflict with human beings and the gain which is fought for is granted, not by nature, but by man [sic]' (42). Urban social relations are thus constituted by conflict, or more accurately, competition among increasingly atomized individuals over money, commodities and space in the *absence of nature*. Thus, in Simmel's depiction nature is irrelevant to the metropolis; both in concrete struggles over 'gain' and in exchange relations. For Simmel the city is beyond and apart from 'nature'.

Louis Wirth's 'Urbanism as a way of life' (1938) is less frightful in tone, and even celebrates 'the urbanization of the world' as 'one of the most impressive facts of modern times' (1). His modernist outlook, however, does not cloud his judgment of what the urban way of life lacks: 'nowhere has mankind been farther removed from organic nature than under the conditions of life characteristic of great cities' (1–2). For Wirth, this might be a positive attribute of cities, but he reaffirms Simmel's erasure of nature from the urban. In order to understand what this lack of nature might mean in Wirth's assessment of 'urbanism' I move on to unpack his theories on relations between people.

Wirth argues that the combination of population size, density, and heterogeneity in urban areas produces a particular condition of 'urbanism'. He suggests that face-to-face primary relations of the rural past are supplanted by fleeting and impersonal interactions of the urban present. Like Simmel, Wirth highlights the superficiality of social interaction in the city. Leaving aside the empirical validity of Wirth's characterization of urbanism (cf. Guterman 1969, for discussion), I contend that Wirth's conception of 'primary' social relations is analogous to society–nature relations. As Wirth describes:

> Characteristically, urbanites meet one another in highly segmental roles ... This is essentially what is meant by saying that the city is characterized by secondary rather than primary contacts. The contacts of the city may indeed be face to face, but they are nevertheless impersonal, superficial, transitory, and segmental. (1938: 12)

For Wirth, primary contacts are more authentically social – based on longstanding relations of family, friendship and social support – but, secondary contacts, are 'superficial'. Yet, this dualistic conception does not explain the deep sociality of secondary contacts. The transitory relations of monetary exchange in the city bring into relation global geographies of commodity production, transportation, and distribution. Likewise, Wirth's description of mere 'secondary' social contacts can be extended to the world of 'organic nature', which in the built urban environment putatively exists in processed, 'secondary' forms. The urban imaginary of nature can only conceive of a 'primary' engagement with nature either in the direct production of one's livelihood from the land, or in a more romantic spiritual communion with natural areas. Yet, however transitory and superficial, everyday urban practices are deeply entangled with global geographies of resource production. Indeed, urbanism can be also understood as lacking 'primary contacts' with 'organic' nature, as Simmel discussed above.

Since cities are the centres of capitalist exchange, they are concurrently centres of fetishism (Marx, 1976 [1867]), where the links between nature and community are seemingly lost. Simmel understood that these exchange relations are highly intensified in cities, but he did not consider that exchange simply fetishizes its real social and ecological basis. Wirth both eschews the sociality of the secondary contacts and displaces nature from the city. Both Simmel and Wirth's accounts thus reify such alienation and acquiesce to the *apparent* absence of nature and primary social relations in the capitalist metropolis. Indeed, part of what makes 'capitalist urbanization' capitalist is that the majority of people living in the city do not possess any means of producing their own livelihood without the wage-relation. The capitalist city, therefore, presupposes that the majority of individuals are, first, alienated from the results of their labour and, second, dependent on Simmel's world of exchange relations vis-à-vis the wage-relation for their most basic needs.

On the one hand, this urban imaginary of nature in opposition to the urban carves out a whole realm of politics conceived purely as 'urban'. For example, struggles over housing, transportation, homelessness, or other 'urban' issues do not tend to confront the ecological entanglements of such infrastructures. More importantly, the urban imaginary of nature allows for the emergence of a specifically urban politics of a narrowly conceived 'nature'. Simmel and Wirth's alienating conception of urban life, where cities and humans are seem as separated from nature, presents political opportunities for various social forces seeking to reconcile the conflictual aspects of the metropolis. In Simmel's case, the conditions producing a particular type of blasé, monetized mental life hasten counter-imaginaries seeking more communal, cooperative social relations not only between humans but also with the world of nature that has been abstracted and reduced to money. For Wirth, the lack of 'primary contacts' (i.e. meaningful relations of interdependence) will produce counter-imaginaries constructing ideals of more authentic, primary social relations. The seeming absence of nature in cities overflowing with fetishized commodity relations facilitates the construction of an ideal of nature. Simmel's notion of the urban helps explain the romantic construction of nature as *something external to cities* (Cronon, 1995) – some un-monetized, lost land that we must recover and reconnect with (Cronon, 1995; Merchant, 1995) – while Wirth's 'organic nature' gets reified as a *thing*, external not only to the individual urbanite, but to the entire urban spatial form itself. Such opposition (urban/nature), if considered dialectically, discloses their mutual constitution and moreover reveals 'nature' as mere antithetical reflections of Simmel and Wirth's particular conception of the urban.

> **BOX 12.1 THE CHANGING NATURE OF AMERICAN ENVIRONMENTAL POLITICS**
>
> Since urbanization is constructed as a denaturalized process, political forces have emerged to reinsert a particular vision of 'nature' back into cities. Beginning in the nineteenth century, urban landscape architects like Frederick Law Olmsted believed that cities needed to construct parks as specific 'nature' zones (complete with trees, gardens, lakes and open fields) for populations to recuperate from the unnatural and alienating pace of urban life. Today, there still exists a strong 'open space' movement that attempts to quarantine natural 'green' spaces from the impending threat of the closed spaces of expanding suburban sprawl.

The absence of material nature upon which humans can labour creates a specifically *urban* imaginary of nature, where cities often stand in as a contradictory antithesis to a construction of *pristine* nature (e.g. wilderness spaces). The urban erasure of an imagined 'nature' creates the space for a specific kind of politics which calls for the reclamation and reunification of society with a 'nature' imagined in opposition to the city. Of course, recent literature under the banner of 'urban political ecology' has served to make visible the socio-natural relations that actively constitute what we construct as 'urban' (Heynen et al., 2006), but this does not necessarily prevent the continuation of cultural political constructions of 'nature' in opposition to the urban. In what follows, I will consider three key texts that inform mainstream environmental politics in the United States to show their unique constructions of nature in relation to cities.

The Urban Imaginary of Nature and Environmental Politics in the United States

Sand County Almanac: The 'Biotic Community' and the Violence of Urbanization

Aldo Leopold's *Sand County Almanac* (1949) was a landmark text in the conservation movement in that it was able to translate the ideas of ecological science into popular forms (Gottlieb, 2005: 67–70). Drawing from his own observations of the non-urban wilderness landscapes around his home in rural Wisconsin and around various wilderness areas in the United States, Leopold illustrated the complex set of ecological relationships that constituted what he called the 'biotic community' (Leopold, 1949: 191). The biotic community was understood as a set of internal organic relations between multiple species that combine to provide coherence and stability to living systems over long periods. Out of this specific appreciation of 'land' as a 'community' (xviii) based upon harmonious, mutually beneficial set of ecological relations, came Leopold's politics of the 'land ethic': 'A thing is right when it tends to preserve the integrity, stability, and beauty of the biotic community. It is wrong when it tends to do otherwise' (262).

Insofar as the text is a celebration of wilderness, much of the book does not consider the city at all. However, Leopold offers sharp critical asides about the need to find refuge in nature from 'too much modernity' (xviii). In his history of the landscape as reconstructed through the tree rings of a single Wisconsin 'good oak' (6), he discusses the 1890s as a period of ecological destruction, which was celebrated for its prosperity and economic growth, 'by those whose eyes turn cityward rather than landward' (13). Perhaps his strongest critique was for the emerging 'recreational' use of nature (camping, fishing and hunting) that served to degrade rather than appreciate wilderness spaces in their own right. He was clear on the source of this degradation: 'Recreation became a problem … when the railroads … began to carry city-dwellers, *en masse*, to the countryside' (280). The problem with city-dwellers, or later 'motorized tourists' (281) was their lack of ecological conscience necessary for appreciation of the 'biotic community'. For the hunter, for instance, the goal is not the appreciation of the *multiple* species that make up this community, but rather, the 'trophy' of the single species that the hunter collected while in nature: 'It attests that its owner has been somewhere and done something' (284).

BOX 12.2 URBAN POLITICAL ECOLOGY AND THE METABOLIC CITY

Urban political ecology attempts to deconstruct the dualism between cities and nature by emphasizing the socio-ecological relations that underpin every aspect of urban life. From such a perspective comes the famous statement by David Harvey (1996: 186): 'there is nothing unnatural about New York City'. For example, several urban political ecologists have not only shown the necessary relation between urban inequality and the uneven distribution of water resources, but also traced the geographically dispersed 'hydrosocial' networks necessary to bring water in and out of cities (see, e.g., Gandy, 2002; Swyngedouw, 2004; Kaika, 2005).

Central to the ethical politics of the 'biotic community' is of course the 'perfect norm ... [of] ... wilderness' that has 'maintained itself for immensely long periods' (274). In one passage in the famous 'land ethic' chapter, it becomes clear that Leopold believes that cities themselves – characterized by Wirth's notion of population density – are *by definition* spaces of the violent destruction of nature. Leopold begins by asserting what later became a dictum of specifically Malthusian environmentalist discourse of 'carrying capacity' (cf., Harvey, 1974; Sayre, 2008): 'Many biotas ... have already exceeded their sustained carrying capacity.' He goes on to conclude that '[M]ost of South America is overpopulated' (257). Leopold expands on the logic of ecological 'violence':

> The combined evidence of history and ecology seems to support one general deduction: the less violent the man-made changes, the greater the probability of successful readjustment in the pyramid. *Violence, in turn, varies with human population density; a dense population requires a more violent conversion* ... All gains from density are subject to a law of diminishing returns. (257–258; emphasis added)

Only though this logic is it made clear that Leopold's 'land ethic' – that is to say Leopold's *politics* – cannot integrate densely populated cities, and their inhabitants, as proper political subjects in the construction of ecological harmony between nature and society. If cities are centres of modernity then 'your true modern is separated from the land by too many middlemen and by innumerable physical gadgets' (261). Indeed, in this view, the city itself not only *presupposes* a level of destruction of the 'biotic community', but also a disconnection from a certain imagined 'nature' that was – following Simmel and Wirth – external to cities.

To be sure, twentieth-century processes of urbanization were violent and destructive to the world of non-human nature (McNeill, 2000). However, it is only through this seemingly unnatural process of urbanization that the imaginary of 'the perfect norm of wilderness' is created. Thus Leopold's imaginary of the 'biotic community' emerges from the historical experience of urbanization. As Cronon (1995) has argued, wilderness is the imagined 'other' – the antithesis – of the city. Thus, out of Leopold's writings emerged a specific form of environmental politics that sought to protect – and spatially delimit – natural wilderness spaces from the ravages of the expanding city. Yet, how cities themselves could become political subjects in effort to reconcile the destructive capitalist relations between society and nature was, at best, left ambiguous, and, at worst, painfully clear; the reunification of humans with the 'biotic community' would be *impossible* alongside the existence of cities. Such a scenario suggests only the possibility of a paradoxically *urban* environmental politics against cities.

Silent Spring: Toxic Chemicals and the 'Repository Environment'

Rachel Carson's *Silent Spring* (1962) is usually credited with the rise of the environmental movement during the late 1960s (Gottlieb, 2005: 121–127). While Leopold, John Muir, Gifford Pinchot and other icons of the conservation and/or preservation movements advocated for a kind of spatial politics of *protection* Carson's politics raised public consciousness around a new form of environmental menace: synthetic chemicals and most famously the insecticide DDT. In painstaking detail, Carson reviewed how the post-World War II proliferation of these 'elixirs of death' (15) had systematically poisoned ecological systems: from bird populations, to agricultural soils; from river systems, to human bodies themselves. In one sense, Carson's notion of the ecological interrelationships of nature is similar to Leopold's: 'The earth's vegetation is part of a web of life in which there are intimate and essential relations between plants and the earth, between plants and other plants, between plants and animals' (64). On the other hand, Carson's work transformed the vision of nature structuring environmental politics. Rather than a 'nature' in need of spatial protection from resource-based industries or haphazard recreational nature tourism, Carson called attention to the role of nature as a 'sink' for not just chemicals, but all forms of industrial pollution (Jones, 2002). Thus, Carson constructed nature as a kind of 'repository environment' for poisonous pollution that kills biological life (including humans) and destroys the complex harmony of ecological systems.

Like Leopold, Carson's narrative is striking for the *absence* of cities or any discussion of urban systems. Much of the analysis is focused on rural agricultural areas and forests threatened by insects whose destructive power was countered with the mass spraying of DDT and other petrochemical pesticides. Carson beautifully traces the *circulation* of these chemicals through the dense webs of ecological exchange far beyond the initially sprayed farms and forests. If cities are spoken of, it is often as sites of knowledge production. Carson constantly relies on evidence from authoritative centres or universities in cities whose practitioners had conducted studies proving the deadly consequences of chemical pollution (e.g. The Wildlife Research Unit in Auburn, Alabama (164); The Sloan Kettering Institute in New York (233); Royal Victoria Hospital in Montreal (236)). In fact Carson ridicules spraying programmes in New York City that attempted to eliminate the pesky gypsy moth: 'The gypsy moth is a forest insect, certainly not an inhabitant of cities' (158).

In Carson's narrative, chemicals emanate outwards from human settlements like farms, forest conservation zones, and even suburban homes with their gardens and household cleaning products. The only escape from the 'sea of carcinogens' (239) attached to chemical pollution is isolation from human settlement:

> Probably no person is immune to contact with this spreading contamination unless he [sic] lives in the most isolated situation imaginable ... To find a diet free from DDT and related and chemicals, it seems one must got to a remote and primitive land, still lacking the amenities of civilization. (174, 179)

Anticipating the *global* nature of environmental pollution problems, Carson relays one such imaginable example of isolation – Alaskan Eskimos. But even they have not escaped:

> When some of the Eskimos were checked by analysis of fat samples, small residues of DDT were found ... The reason for this was clear. The fat samples were taken from people who had left their native village to enter the United States Public Health Service Hospital in Anchorage for surgery. There the ways of civilization prevailed, and the meals in this hospital were found to contain as much DDT as the most populous city. For this brief stay in civilization the Eskimos were rewarded with the taint of poison. (179–180)

Thus, a mere visit to the city of Anchorage made it more likely to encounter the 'poisons' of civilization. Moreover, civilization is at the centre of chemical circulation and, thus, for the Eskimos, the only way to escape the 'taint of poison' would have been to remain in their villages and never visit Anchorage. By 'civilization' Carson obviously invokes *urbanization*, or what she also calls, 'our modern way of life' (188). Like Leopold, and Wirth, it is this *urban* way of life that is characterized by a profound *disconnection* and disregard for nature: 'Most of us walk unseeing through the world, unaware alike of its beauties, its wonders, and the strange and sometimes terrible intensity of the lives that are being lived about us' (249). For city dwellers the fate of the 'repository environment' – as evidenced by the 'silent spring' of killed bird populations – is invisible and unacknowledged, unless it becomes clear that the repository is to become transformed from the non-human 'environment' to human bodies themselves. Thus, for Carson the object of political concern extends beyond the non-urban spaces of wilderness and 'nature' that concern Leopold to focus upon human bodies, cities and any 'sink' that absorbs the circulation of chemicals and pollutants. Yet, she shares Leopold's focus on urban civilization as the main driver of ecological destruction, and, therefore, reinforces an anti-urban environmental politics.

An Inconvenient Truth: Becoming Political Through Escape from the 'Urban Frenzy'[3]

Al Gore's *An Inconvenient Truth* (2006) represents the popular apotheosis in yet another stage in the evolution of the American environmental movement. After the 1970s and 1980s were characterized by the politics of the 'repository environment', a new form of *global* environmental politics concerned itself with the biospheric consequences of pollution (Mansfield, 2008). The 1990s were characterized by a diverse set of global environmental concerns including ozone layer depletion, rapid biodiversity loss and deforestation. Yet, the concern of global warming, or what was later revised as 'climate change', steadily rose to subsume them all. By the mid-2000s as the scientific evidence and more everyday experiences of hotter temperatures and stronger weather events increased global public consciousness on the possibility of catastrophic climate change marked by rising sea levels, desertification, longer heat waves, and the possibility of human extinction (Davis, 2010). As evidenced by the 2007 Nobel Peace Prize, Al Gore's *An Inconvenient Truth* is comparable to Carson's *Silent Spring* in that it raised consciousness and a sense of political urgency around the issue of climate change. The film and the text are shot through with images of urbanization as a core driver of climate change. First of all, the classic imagery of 'greenhouse gases' is constructed as the *infrastructure* of urban life. The book includes double page images of a (presumably coal-fired) electric power plant spewing carbon dioxide into air (24–25) and the congested inter-state highway peppered with automobiles (28–29). These are the energetic bases of everyday urban existence. Like Leopold, the spectre of population growth is posited as a specifically urban problem. Alongside an aerial photo of an endless landscape of skyscrapers and urban settlement in Tokyo,[4] the text reads: 'Most of the increase in population is in developing nations where most of the world's poverty is concentrated... and most of the increase is in cities' (218–219). However, perhaps the most powerful imagery of the film and book is the time-scaled graph that shows the increasing concentration of CO_2 in the atmosphere. Famously in the film, Gore is raised high into the air with a mechanical step-stool to be able to physically point to the skyrocketing elevation of CO_2 over the last 50 years. The root origins of rising CO_2 begin with the expansion of urbanized industrialization in the nineteenth century: 'Pre-industrial concentration of CO_2 was 280. In 2005, that level, measured high above Mauna Loa was 381 parts per million' (37). Thus, the problem of climate change is indisputably rooted in the experience of a specifically industrialized form of urbanization and its reliance upon fossil fuel energy for the basics of capital circulation and everyday social reproduction (Huber, 2009).

Apart from Gore's presentation of the problem of climate change itself, he also conveys a remarkably old-fashioned account of his own personal political transformation toward environmental concerns. In Gore's view, this political transformation was not only rooted in education (and, of course, his mother reading him *Silent Spring* as a boy; 124), but escape from urbanized environments. In a section aptly titled 'Concrete and Countryside', he discusses his family trips outside Washington, DC to his family's '[B]ig sprawling, beautiful farm in Tennessee with animals, sunlight and grass, all nestled in a sweeping bend of clear and sparkling Caney Fork River' (122). It was there, where Gore was able to discover a lost connection with the natural world. Gore's account sounds very similar to Theodore Roosevelt's construction of nature as a site of refuge and rejuvenation of masculinized virility (Haraway 1984): 'I breathed freely – full chested, invigorating breaths – unlike any I ever took on the streets of Washington, DC' (122). Gore's celebration of recreational nature enthusiasm even lacks the critical perspective of Leopold over a half-century earlier. Only through family camping and hiking trips in non-urban locations, could Gore 'shake off the urban frenzy …', 'spend unscheduled time in untame places,' and find a 'sense of renewal' (160). Like Simmel and Wirth, Gore's conception of the urban was a space in the absence of nature: 'It's such an encompassing artificial environment that it can seem to be all there is' (161). Thus, Gore's own process of *becoming political* about climate change and environmental politics necessarily was in opposition to the 'artificial' city as an unnatural site of ecological despoliation and desensitization from ecological relationships.

Finally, near the end, Gore recommends a set of solutions available to individuals to confront the climate crisis. Like Carson and Leopold, the problem of environmental politics is specifically an urbanized 'way of life': 'The truth about the climate crisis is an inconvenient one that means we are going to have to change the way we live our lives' (286). The object of political activity is a set of urbanized practices of 'exchange', highlighted as characteristic of the mental life of the metropolis by Simmel. Alongside the image of a suburban home (with its implications of property ownership and a certain level of material consumption), the text suggests purchasing new appliances, light bulbs, having your house audited, driving smarter, 'modifying your diet to include less [but not eliminate] meat', and considering 'the impact of your investments' [this of course assumes your positionality as a member of the investor class] (310, 311, 317, 322). Characteristic of the lack of 'primary' social relations in the city as highlighted by Wirth, the only political option available it seems is mostly individualized practices of monetized consumption. Given the subsumption of this form of politics within the neoliberal logics of 'exchange' and decentralized 'individual' decision-making, it should come as no surprise that Gore's main policy proposal is a market-based cap and trade system that seeks to use 'market capitalism as an ally' (270). Using the logics of the market, Gore believes urban life can be reconfigured toward greener futures, but left unresolved is how market logics can solve the artificiality of exchange-centred metropolises.

Conclusion

Mike Davis (2010) recently proclaimed that the 'single greatest cause of global warming – the urbanization of humanity – is also potentially the principal solution to the problem of human survival in the later 21st century' (30). The above analysis illustrates the difficulty of mainstream American environmental politics envisioning the city as a *solution* to global ecological crisis. As Williams (1980: 67) famously put it, '[the idea of] nature ... contains an extraordinary amount of human history'. In this chapter, I have drawn attention to the historical constitution of ideas of 'nature' and, moreover, situate such ideas at the centre of processes, like urbanization, that are constructed as most unnatural. At its core, the urban imaginary of nature can be seen as reflective of the clustering of fetishized exchange relations in cities, masking the real dependence of the city on social relations between labour and the material world.

What I have tried to demonstrate specifically is that the urban construction of 'nature' is a recurrent theme within the history of American environmental politics. Whether expressed as the conservationist 'land ethic,' 'repository environment', or 'climate crisis,' urbanization is not only constructed as the core driver of the violent destruction of nature, but also an unnatural site marked by a desensitized 'blasé outlook' towards the complexity (and beauty) of ecological relationships. Although not expressed explicitly, this particular imaginary of nature points paradoxically to an *urban* environmental politics *against* cities. While I want to emphasize the political possibility in oppositional ideas such as 'nature', I also want to point out that this politics is not radical enough. The idea of nature itself often serves to obfuscate the socio-ecological relations that make the city what it is. Thus, while it may offer protection to certain naturalized spaces (forests, farmlands, rivers) it forgets the environmental justice and sustainability problems that afflict denaturalized urban spaces.

Since the urban imaginary of nature is in opposition to the city, it allows for, on the one hand, a vast field seen as purely urban politics (e.g. housing, transportation, culture) and, on the other hand, a rather narrow set of concerns constructed as 'environmental' (e.g. parks, urban agriculture, forests). Thus, while this chapter is a critical account of the politics of the urban imaginary of nature, a more radical urban environmental politics might consider a more *denaturalized* urban environmental politics (Biro, 2005) focused on what urban political ecology scholars call the 'politics of the urban metabolism' (Heynen et al., 2006). By focusing on the middle spaces of exchange and flows, the idea of 'metabolism' belies any dualism between cities and nature. For example, an urban politics of climate change cannot simply focus on those naturalized spaces in need of protection from the encroaching city (e.g. protecting urban forest cover for its capacity to absorb CO_2), but must also confront those spaces constructed as 'unnatural' such as electricity grids, transportation networks and built environments. The increasing concern over climate change suggests such an urban environmental politics is emerging (Bulkeley and Betsill, 2003). Such a politics would not only point vaguely to the unsustainability of modern ways of life, but also point out how these ways of life are reproduced through uneven relations of power and injustice over access to and control over the infrastructures of the urban metabolism. Thus, city-dwellers themselves can become political subjects in the efforts to reconstruct just and sustainable socio-ecological futures. Such a politics, however, would depend on the proliferation of metabolic 'imaginaries' that – like the urban imaginary of nature – provide the discursive terrain for forging broad political coalitions across a variety of social interests. The question therefore is: can an urban imaginary of the metabolism be as powerful as the imaginary of nature?

Notes

1. Sections of this chapter are reprinted with permission from *Urban Geography* (Huber and Currie, 2007), Vol. 28, No. 8, pp. 705–731. © Bellwether Publishing, Ltd., 8640 Guilford Road, Columbia, MD 21046. All rights reserved.
2. This should not be confused with what Gandy (2006) calls the urban ecological imaginary referring to the mechanistic ecological metaphors used by the Chicago School.
3. For this section I rely on the published book *An Inconvenient Truth* (2006). This book is very similar to the popular and influential film, and its textual form of presentation is useful for my purposes.
4. Given the attached text, it's curious that the image is of Tokyo, which is hardly the epicentre of the urban population explosion in the so-called 'developing world'.

References

Bosso, C. and Guber, D. (2003) 'The boundaries and contours of American environmental activism', in Vig, N. and Craft, M. (eds) *Environmental Policy: New Directions for the 21st Century*. CQ Press: Washington, DC. pp. 79–102.

Biro, Andrew (2005) *Denaturalizing Ecological Politics: Alienation from Nature from Rousseau to the Frankfurt School and Beyond*. University of Toronto Press: Toronto.

Brenner, N. (2000) 'The urban question as a scale question: Reflections on Henri Lefèbvre, urban theory, and the politics of scale', *International Journal of Urban and Regional Research*, 24(2): 361–378.

Bulkeley, H. and Betsill, M. (2003) *Cities and Climate Change: Urban Sustainability and Global Environmental Governance*. Routledge: London.

Carson, Rachel (1962) *Silent Spring*. Houghton Mifflin: Boston, MA.

Castells, Manuel (1977) *The Urban Question: A Marxist Approach*. Arnold: London.

Castree, N. and Braun, B. (1998) 'The construction of nature and the nature of construction: analytical and political tools for building a survivable future', in Braun, B. and Castree, N. (eds) *Remaking Reality: Nature at the Millennium*. Routledge: London. pp. 3–42.

Cronon, W. (1995) 'The trouble with wilderness; or getting back to the wrong nature', in Cronon, W. (ed.) *Uncommon Ground: Rethinking the Human Place in Nature*. W.W. Norton: London. pp. 69–90.

Davis, M. (2010) 'Who will build the ark?', *New Left Review*, 61 (January–February): 29–46.

Demeritt, D. (2002) 'What is the "social construction of nature"? A typology and sympathetic critique', *Progress in Human Geography*, 26(6): 766–789.

Dowie, M. (1995) *Losing Ground: American Environmentalism at the Close of the Twentieth Century*. Cambridge, MA: MIT Press.

Fitzsimmons, M. (1989) 'The matter of nature', *Antipode*, 21(2): 106–120.

Foucault, Michel (1978) *The History of Sexuality Vol. 1*. Pantheon Books: New York.

Gandy, Matthew (2002) *Concrete and Clay: Reworking Nature in New York City*. MIT Press: Cambridge, MA.

Gandy, M. (2006) 'Urban nature and the ecological imaginary', in Heynen, N., Kaika, M. and Swyngedouw, E. (eds) *In the Nature of Cities: Urban Political Ecology and the Politics of the Urban Metabolism*. Routledge: New York. pp. 63–74.

Gore, Al (2006) *An Inconvenient Truth: The Planetary Emergency of Global Warming and What We Can Do About It*. Rodale: Emmaus, PA.

Gottlieb, Robert (2005) *Forcing the Spring: The Transformation of the American Environmental Movement*. Island Press: Washington, DC.

Guterman, S. (1969) 'In defense of Wirth's "Urbanism as a way of life."', *American Journal of Sociology*, 74: 492–499.

Haraway, D. (1984) 'Teddy bear patriarchy: Taxidermy in the Garden of Eden, New York City, 1908–1936', *Social Text*, 11 (Winter 1984–1985): 20–64.

Harvey, D. (1974) 'Population, resources, and the ideology of science', *Economic Geography*, 50(3): 256–277.

Harvey, David (1996) *Justice, Nature and the Geography of Difference*. Blackwell: Oxford.
Hays, Samuel (2000) *A History of Environmental Politics since 1945*. University of Pittsburgh Press: Pittsburgh, PA.
Heynen, Nik, Kaika, Maria and Swyngedouw, Erik (eds) (2006) *In the Nature of Cities: Urban Political Ecology and the Politics of Metabolism*. Routledge: London.
Huber, M. (2009) 'Energizing historical materialism: Fossil fuels, space and the capitalist mode of production', *Geoforum*, 40(1): 105–115.
Huber, M. and Currie, T. (2007) 'The urbanization of an idea: Imagining nature through Urban Growth Boundary Policy in Portland, Oregon', *Urban Geography*, 28(8): 705–731.
Jacobs, Jane (1961) *The Death and life of Great American Cities*. Random House: New York.
Jones, D. (2002) 'The earth as output: Pollution', in Johnston, R., Taylor, P. and Watts, M. (eds) *Geographies of Global Change: Remapping the World*. Blackwell: Oxford. pp. 391–411.
Kaika, Maria (2005) *City of Flows: Modernity, Nature, and the City*. Routledge: London.
Leopold, Aldo (1949) *A Sand County Almanac*. Oxford University Press: New York.
Low, Nicholas (2000) *Consuming Cities: The Urban Environment and the Global Economy*. New York: Routledge.
McNeill, John (2000) *Something New Under the Sun: An Environmental History of the Twentieth-Century World*. W.W. Norton and Co: New York.
Macek, Steve (2006) *Urban Nightmares: The Media, the Right, and the Moral Panic Over the City*. University of Minnesota Press: Minneapolis, MN.
Mansfield, B. (2008) 'Global environmental politics', in Cox, K. (ed.) *The Sage Handbook of Political Geography*. Sage: Thousand Oaks, CA. pp. 235–246.
Marx, Karl (1976) [1867] *Capital Vol. I*. Translated by B. Fowkes. Vintage Books: New York.
Merchant, C. (1995) 'Reinventing Eden: Western culture as a recovery narrative', in Cronon, W. (ed.) *Uncommon Ground: Rethinking the Human Place in Nature*. W.W & Norton Co.: New York. pp. 132–170.
Peet, R. and Watts, M. (1996) 'Liberation Ecology: development, sustainability, and environment in the age of market triumphalism', in R. Peet and M. Watts (eds) *Liberation Ecologies: Environment, Development and Social Movements*. Routledge: London. pp. 1–45.
Rome, A. (2001) *The Bulldozer in the Countryside: Suburban Sprawl and the Rise of American Environmentalism*. Cambridge University Press: New York.

Sayre, N. (2008) 'The genesis, history, and limits of carrying capacity', *Annals of the Association of American Geographers*, 98(1): 120–134.

Simmel, G. (1995) [1903] 'The metropolis and mental life', in Kasinitz, P. (ed.) *Metropolis: Center and Symbol for Our Times*. New York University Press: New York. pp. 30–45.

Swyngedouw, E. (2004) *Social Power and the Urbanization of Water: Flows of Power*. Oxford University Press: New York.

Swyngedouw, E. (2006) 'Metabolic urbanization: the making of cyborg cities', in Heynen, N., Kaika, M. and Swyngedouw, E. (eds) *In the Nature of Cities: Urban Political Ecology and the Politics of Urban Metabolism*. Routledge: London. pp. 21–40.

Williams, R. (1980) *Problems in Materialism and Culture*. Verso: London.

Wilson, A. (1991) *The Culture of Nature: North American Landscapes from Disney to Exxon-Valdez*. Between the Lines: Toronto.

Wirth, L. (1938) 'Urbanism as a way of life', *American Journal of Sociology*, 44: 1–24.

Zukin, S., Baskerville, R., Greenberg, M., Guthreau, C., Halley, J., Halling, M, Lawler, K., Neiro, R., Stack, R., Vitale, A. and Wissinger, B. (1998) 'From Coney Island to Las Vegas in the urban imaginary: Discursive practices of growth and decline', *Urban Affairs Review*, 33(5): 627–654.

13

URBAN POLITICS AS PARALLAX

Deborah Martin and Mark Davidson

We have set out in this book to reconsider urban politics, as a topic of urban geographical scholarship. Central to our approach is the multidimensionality of such politics. In our introduction we argued, following Žižek, that a critical eye on urban politics would not only acknowledge but also seek a parallax view: a viewpoint on the topic that is beside, beyond, and outside the first, primary, obvious, and perhaps default and customary perspective(s). Such parallax imagery is useful because it reminds us that to see another viewpoint, we sometimes have to completely suspend the viewpoint we already have. Yet even the parallax is limiting, since it implies a double-sided image. The chapters in this volume together point to the ways that our views on urban politics cannot be even two-sided; but instead are multi-dimensional. They collectively highlight the importance of taking a viewpoint in order to see something, and encourage our exploration of alternative lenses for seeing and investigating urban politics.

At the outset of the book, we presented three entry positions to urban politics – that of setting, medium, and community – which offer different pathways to seeking the view that isn't obvious. As we unpack here, however, from that starting point, our contributors point to further shifts in view that prompt new perspectives and understandings of urban politics. What is important to understand in all of this shifting is that the questions that we start with, and the assumptions that we make about where we find urban politics, will shape our subsequent understandings of it. The key point is therefore that we require a process of continual critical reflection upon the questions of what is urban and what is political in order that we make sense of the actors, relations, and places that comprise and condition cities and citizens. Only with such a process in place can we avoid the trappings of a singular or limited set of viewpoints that might themselves serve to limit the potential for political change.

A primary theme for our examination of urban politics comes out of the very first set of chapters, and is echoed again in subsequent ones. This theme is that, even as a setting for politics, urban politics cannot be adequately conceptualized as contained in a singular place-that-is-urban. In the first section, the chapters by Katherine Hankins and Deborah Martin, Kevin Ward, and Kathe Newman and Elvin Wyly each point to the ways that urban politics are simultaneously experienced, practised, and waged over specific conditions of the city – local concerns like housing, garbage collection, and land-use planning – and yet these local issues are always bound up in conditions and relations of multiple locations and scales. The fixity or boundedness of a thing called urban politics is immediately called into question in these chapters, and in later chapters like Mark Davidson's and Matt Huber's. The authors do not seek to do away with conceptualizing urban politics, but they insist that investigation of particular cases of a local requires looking beyond, seeking out the ways that seemingly grounded neighbourhood or city-specific planning lies in relationships, and processes far beyond the city limits, that necessarily shape the experience of the local and the politics of making 'local' choices.

The chapters by Hankins and Martin, on neighbourhoods as a site for urban politics, and Newman and Wyly, on the localization of global mortgage finance, perhaps represent polar opposites in the question of urban-as-setting for politics. Yet a closer look reveals the ways that urban and global, local and relational, mutually and fluidly co-constitute through power relations and decision-making, or ordering and categorizing of urban land and its uses. In the first, the setting is unabashedly local urban consumptive practices of city residents: neighbourly relations, housing acquisition, and urban services like garbage collection (neighbourhood as the site of collective consumption). In such a view, neighbourhood politics that enrol the local state in (improving) service provision are indeed a hallmark of urban politics. Hankins and Martin interrupt this primarily local view of neighbourhood, however, to consider how the conditions of any neighbourhood come from its situation in relation to processes within and far beyond it; especially in the case of the poor inner-city neighbourhoods that are the sites of their study of 'strategic neighbouring'. Their conclusion, that politics are only contained and assertive of a structural (organizing) status quo when they are situated in neighbourhoods, evokes the very processes that Newman and Wyly unpack; processes of financial markets that, far from being aspatial and inattentive to local place, respond to and re-inscribe the conditions of place. As Newman and Wyly state in relation to understanding the conditions of poor neighbourhoods:

> [T]he problems within communities look local, fragmented, and caused by the people in these places. An urban politics of finance can tie these local experiences together to systematically understand the policy choices made that constructed a political economy in which the foreclosure crisis was possible.

The significance of this parallax view of urban politics as not really neighbourhood, but relational, simultaneously setting and process lies in the examination of actors, processes and playing out; that is, in the ways that specific processes or relations find material expression in urban places with consequences for the social and political relations both here and there, urban and regional/national/global. The seemingly local politics of place, then, are indeed urban politics, ones that evoke attention to processes of place production and neighbourhood differentiation.

Our second theme emerges in the interstices of Section 1 'Setting' and Section 2 'Medium', as the chapters shift to the power relations and dynamics of politics. Here, the theme is how the formal structures of government situate urban politics as medium – a means to enact municipal policies – even as they enrol broader processes and shape what is accepted as local and political. As chapters by Kevin Ward, Kurt Iveson, Donald McNeill, and John Carr demonstrate, the formal setting of local government is not only means for power, but a medium through which power and decisions are framed – an enforcing mechanism for understanding and shaping 'urban' and 'politics'. As Kevin Ward illustrates, the operations of politics entail intersecting and coalescing processes into place (local politics) and back again into processes across and outside urban governance. Ward contrasts two different urban politics, both of which, as in the chapters by Hankins and Martin, and Wyly and Newman, see the urban as a setting for questions of urban service provision, and economic development. In Ward's chapter, however, the purview of formal urban government defines the setting of politics, as well as the issues, even when they invoke broader processes like capital investment and economic restructuring. Indeed, even as Ward introduces the idea of splintered urban politics – picked-up policy fragments from cities near and far, reworked through a range of networks – the urban remains primarily about the invocation of power by local government, albeit through multiscalar relations of individual policy-makers.

Combining Ward's viewpoint with those of Donald McNeill and John Carr, we see an emphasis on actors such as elected and appointed officials and their interactions with, or on behalf of, urban citizens. Through the lens of formal governance, two processes are evident. First, local governments comprise influences near and far, shaped not just by local dynamics, personalities, and power structures, but by the ways that such personalities derive ideas, and support from a geographically disparate range of networks. Second, where policy frameworks explicitly incorporate local and grassroots input in the formal setting of local government, these bottom-up insights always highlight formal decision-making structures; according to Carr, they 'occlude' the conflicts inherent in participation in favour of consensus. Carr's discussion of participatory planning in Seattle investigates the ways that elites obscure their own policy decisions through delegation of consensus-building and seeming policy-management to staff and citizen-input processes. Donald McNeill examines mayors-as-policy-makers as an embodiment of the urban as a site of business, an economic power, and a setting for identity. As a framework for viewing the roles and actions of local government, then, Ward, McNeill, and Carr point to urban politics as an arena for certain kinds of power relations which enact and build upon political personalities, cross-local information sharing, and policy networks to produce an urban medium which is as much constitutive of what can be political as it is a means for politics. Our second theme thus highlights that urban politics proceed through that which they enact; municipal policies and actors use the medium of government to reinforce the fact of municipal government.

Adding Kurt Iveson's chapter to our understanding of urban politics as both setting and medium broadens our lens from formal governance to informal social relations and regulations as means for enforcing norms of behaviour and forms of contestation, even in realms seemingly far removed from city hall. Although quite different empirical realms, Iveson's study of graffiti 'policing' shares with Carr's examination of participatory planning an interest in the broad range of social controls that apply to politics. Iveson and Carr both consider urban politics as a medium for who naturalizes and decides what voices we hear in political debate and even in 'underground' discourses like graffiti. Iveson's chapter emphasizes that multiple actors – not just those in the formal arena of government – seek to silence, to give voice to some and not others. Drawing these two parts of 'setting' and 'medium' together in the book include the chapters by Iveson, and Hankins and Martin, who all draw explicitly upon political theorist Jacques Rancière (1999) in order to examine the formal and informal means of defining and delimiting acceptable interactions and recognizable actors for urban politics.

The attention through Rancière to the marginal, or unheard and unseen, of politics connects to the third section of the book, urban politics as community, and the ways communities are defined and made visible or legitimate in the public sphere. These chapters together articulate a third theme of urban politics as simultaneously making evident, and obscuring, difference and its political significance. The different approaches in the five chapters in the community section evoke long-standing notions of urban politics as about, alternatively, competing interest groups, or difference and community identity (Dahl 1961; Young 1990). They consider the contestations of inter-group conflicts, or identities, and highlight the ways that certain voices, actions, or ecologies remain unseen in formal urban politics. Natalie Oswin's chapter, for example, points to the processes that render as invisible certain ways of being in the city. Her use of queer theory exposes the ways that urban politics of identity, family, and labour draw upon and make visible and normal only certain behaviours and bodies. In order to 'see' these processes of queering as urban politics, it is necessary to connect, as Oswin does, the policy dots of global labour flows, 'creative' economies, and cultural politics, be they be a city-state like Singapore, nation-states, or city governments. Like the chapters by Davidson, McNeill, and Carr, Oswin turns our attention to the use and configuration of institutions of power in urban politics, and the ways that institutional power relates to everyday citizens and urban identities, including by simplifying and hiding differences such as class or family form.

The chapter by Jamie Winders points to how assumptions that we make about urban politics – such as how communities are constituted – shape what is seen and accepted as political participation. Her case study of Latinos in Nashville illustrates exactly what might be lost when one perspective of urban politics – that which recognizes residents based on neighbourhood affiliation – is prioritized over other possible perspectives, such as cultural or social identifications. Much like John Carr's chapter, her case illustrates how government efforts at inclusion and recognition shape a viewpoint that only sees certain dimensions, residents, and territories, precluding alternative viewpoints or politics. By seeking to see the communities unrecognized in Nashville's politics, Winders is able to point to the parts unseen, the alternative residents, territories, and viewpoints. The chapters in this book help us to push beyond city hall or formal interactions in part by looking at institutional frameworks and community definitions. A broad range of scholarship in urban politics highlights the dilemmas and ideals around citizen-state engagements in urban politics (see Box 13.1).

BOX 13.1 DILEMMAS OF CITIZEN–STATE ENGAGEMENTS IN URBAN POLITICS

Urban planners and policy-makers struggle with creating effective ways to involve everyday citizens in the business of urban governance. While some scholars, including authors of chapters in this book, such as John Carr and Jamie Winders, question whether formal inclusion is either really possible or effective, planners have nonetheless explored ways to make their work more broad based. Leonie Sandercock (1998) describes how planning has changed over the last fifty years from a top-down government activity to a community-based model, at least in theory. She encourages planners to radically position themselves as members of communities rather than as employees of government. When community institutions work to engage in activities like planning and policy-making, however, they can function very much like the 'police' sphere described by Jacques Rancière, limiting the possibilities for radical voices to be heard because of the emphasis on form and consensus-building. Community groups that seek to

(Continued)

> *(Continued)*
>
> rework the processes that shape their urban spaces often face economic constraints in what they can do and how they are funded (Lake and Newman, 2002; Martin, 2004). Henri Lefèbvre is famously quoted as having called for the 'right to the city', but the means for people to enact such rights and rework urban space are very much a matter of discussion and debate (Marcuse, 2009).
>
> ## References
>
> Sandercock, L. (1998) 'The death of planning modernist planning: radical praxis for a postmodern age' in M. Douglass and J. Friedmann (eds) *Cities for Citizens: Planning and the Rise of Civil Society in a Global Age*. John Wiley & Sons: New York. pp. 163–184.
> Lake, R. and Newman, K. (2002) 'Differential citizenship in the shadow state', *GeoJournal*, 58(2–3): 109–120.
> Marcuse, P. (2009) 'From critical urban theory to the right to the city', *City*, 13(2): 185–197.
> Martin, D. (2004) 'Non-profit foundations and grassroots organizing: reshaping urban governance', *Professional Geographer*, 56(3): 394–405.
>
> ## Suggested further reading
>
> Marcuse, P., Connolly, J., Novy, J., Olivo, I., Potter, C. and Steil, J. (2009) *Searching for the Just City: Debates in Urban Theory and Practice*. London: Routledge.

Susan Hanson's chapter takes a different approach to political community, seeing volunteering, entrepreneurship, and community-building as sites of urban politics. By shifting her viewpoint on politics, Hanson allows us to interrogate the changes that can occur though business relationships. Indeed, both Hanson and Carr's chapters focus on how everyday politics are conditioned through certain activities and processes – the need and desire to be entrepreneurial and the political engagements that exercises like participatory planning involve. Carr's chapter interrogates the channelling that state structures entail – the activities and inputs that are considered relevant and the ways that regulation or law shape it. As with Iveson's chapter, there is a social element and disciplining to the urban politics that result, be they formal or informal. In contrast to Iveson's focus on police, and Carr's on planning, Hanson's chapter actually reworks – and challenges the Rancièrian viewpoint on – the meaning of politics. She sees the entrepreneurs of her case study as active participants in community politics despite – or perhaps because of – their removal from the formal or 'police' state and their ability to shift relationships, not to bring voices into the public policy realm.

Hanson's argument forces us to reconsider how we view entrepreneurs. She challenges a notion of politics as primarily exerting power, and reconfigures 'business' actions as empowering beyond traditional economic activity. Perhaps more than any other set of chapters, Hanson's paired with Mark Davidson's discussion offer the parallax view. In Hanson's, capitalists – albeit small scale, self-exploiting – create an urban politics of community that is, in many ways, post-industrial, seeming to transcend an urban politics of class. Davidson's chapter challenges the post-industrial interpretation, and positions class as still central to and embedded within urban politics. Taken together, the chapters are parallax; see one viewpoint, then the other, but conceptually both together are difficult to hold. Juxtaposing them as we have challenges a singularity of urban politics, and forces questions about what we seek to see in our urban analyses. Matt Huber's chapter highlights how viewpoints about what is urban or political become entrenched over time and are hard to challenge.

In this book we have therefore sought to disrupt the notion of urban politics with multiplicity. But we are not advocating an 'anything goes' perspective; rather, the pressing issue concerns what position one takes to consider urban politics, and what possible questions such a position demands and allows. We suggest here three guiding principles or themes: that the urban is always constitutive of near and far, multi-scalar and socio-ecological processes; that urban politics contain and frame possible modes of engagement, necessitating constant questioning of the very structures of engagement and debate; and finally, that urban politics consciously deploys and obscures identity, difference, contestation – rendering some conflict visible and others not. In order to grapple with these multiplicities, we advocate a perspective that consciously seeks to destabilize, to see the side as yet unseen.

What this perspective leaves us with is a commitment to destabilize and deconstruct. But yet, and perhaps more importantly, it also means that at times we must make difficult theoretical and political decisions. When pressed into action or necessitated by events, we must act as political agents and in such situations we need to take positions. This means that some urban political problems might require a fix that, at least for some period, adopts a certain viewpoint that bars others. You cannot, for example, hold all of the authors' viewpoints in this book at the same time. So when it comes to urban political problems such as, for example, fiscal crisis or hate crimes, which perspective must you adopt? And in adopting this perspective which ones can you not hold? This is the difficult politics of taking sides, but as we have hopefully demonstrated in this book, the process of taking sides does not and should not mean a forgetting and/or erasure of other perspectives. The parallax perspective requires a constant consideration of the gaps between perspectives and the implications of holding a perspective that creates a gap between you and those who hold other viewpoints.

References

Dahl, Robert (1961) *Who Governs? Democracy and Power in an American City*. Yale University Press: New Haven, CT.
Rancière, Jacques (1999) *Disagreement: Politics and Philosophy*. University of Minnesota Press: Minneapolis, MN.
Young, Iris M. (1990) *Justice and the Politics of Difference*. Princeton University Press: Princeton, NJ.

INDEX

activism, 25–26
actors, 26, 83
aesthetic, 95
African-American, 30, 62
agency, 115
Allen, John, 6
alliances, 46
Amin, Ash, 164, 166–167
anti-democratic, 125
anti-social behaviour, 92
anti-urban, 55
a-political, 114–115
architects, 91
authority, 89

Banksy, 88, 95
Bentham, Jeremy, 80
Blair, Tony, 91, 92, 105, 193
blasé, 208
Bloomberg, Michael, 104, 106
boundaries, 3
boundedness, 224
bracketing, 1–2, 202
bureaucratic, 123
business owners, 173
Butler, Judith, 142

Calgary, Canada, 49
capital, 44
capitalist, 42, 210
carcinogens, 215
Cartesian, 102
Castells, Manuel, 18, 44
Cato Institute, 49–50
census, 63, 200
Chávez, Hugo, 102
Chicago, 102
Cincinnati, Ohio, 35

citizen, 107, 225
citizen input, 115
city hall, 112
civic affairs, 180
civic memory, 109
class antagonism, *see* class conflict
class conflict, 190, 191
climate change, 196, 214
Clinton, Bill, 193
collective consumption, 44
collective provision, 83
Colorado Springs, Colorado, 172–187
Communist, 106
community, 133–137, 160, 225
    garden, 33
comparison, 48
compromise, 125
conflict, 168
connections, 174
consumption, 45
contestation, 157
Cox, Kevin, 43, 46
creative class, *see* Richard Florida
Critchley, Simon, 134–135
Critical Legal Studies, 113–115
critical, 11
cultural diversity, 161

Davis, Mike, 160, 216
debt, 57
democracy, 27, 124, 134
    participatory, 116
    post-, 36
democratic, 9, 115
depoliticize, 120
derivatives, 58
design, 92
Dikeç, Mustafa, 29

dissensus, 29
distanciated geographies, 102
dual state, 44

economic development agencies, 173
economic growth, 173, 212
elected officials, 225
electoral control, 103
electoral geographies, 102
elites, 113, 122
embeddedness, 46
empowerment, 161, 172
entrepreneurial, 3, 47, 134
entrepreneurs, 172, 173
environmental, 137
environmental movement, 206
environmental politics, 206, 214
equal, 96
everyday, 172, 174

factionalism, 105
family-planning, 147
Fainstein, Susan, 165–167
Federal Housing Administration, 57
feminist, 133, 141, 174, 186–187
financialization, 55
Flores, William, 165–167
Florida, Richard, 144–146
foreign workers, 148–149, 187
Foucault, Michel, 8, 80–2, 107

gays and lesbians, 137, 139–142
gay and lesbian studies, 140
gender, 184
gentrification, 26, 160, 198–199
Gesellschaft, see Tonnies, Ferdinand
Giuliani, Rudy, 102, 107–109
global capital, 72
global warming, *see* climate change
globalization, 3, 4, 190
Goldman Sachs, 104
Gore, Al, 214
Gorz, Andre, 191–192
governance, 3, 8, 37, 105, 163
government, 8, 97, 179
governmentality, 8, 108–109

graffiti, 86, 88, 89–97
  writers, 89, 90–92
Greater London Authority, 196
growth machine, 55, 102, 104, 189

Harvey, David, 5, 47, 103, 134
Hays, Samuel, 206
heterodox economics, 58
heteronormative, heteronormativity, 140, 144, 147, 150
heterosexual, 147
homeless, 114, 181
homosexuality, 139
housing, 55

ideology, 49, 193
imaginary, 207, 218
imagined community, 167
immigrant politics, 157
immigrants, 148, 159
individual, 99
industrial city, 18
inequality, 67, 113
  localization of, 67
injustice, 126
institutional, 227
interculturalism, 167
international finance, 21

Johnson, Boris, 198
judicial system, 115

knowledge, 175

labour, 211
  theory of value, 192
Latino, 30, 62, 157, 158, 167
law, 113
legal, 83, 114, 126
lending, 61
Leopold, Aldo, 211
Livingstone, Ken, 102
local, 25, 224
  dependence, 46
  indicators of spatial association, 61
  government, 179

local, *cont.*
  politics, 23
  state, 45
localization, 70
localized effects, 62
London, England, 195, 197

marginalization, 5, 35
Marx, Karl, 133, 202
Massey, Doreen, 5–6, 195
master plan, 120
mayor, 82, 101–110, 106, 109
mayoral governance, 105
mayoral power, 99
media, 158
  landscapes, 88–89, 93
medium, 79
metabolic city, 212
methodology, 201
middle class, 199
militant particularism, 5
miscount, 86
modernity, 212
Molotch, Harvey, *see* growth machine
money economy, 208
moral geographies, 99
morality, 108
mortgage
  -backed securities, 58
  default, 57
  foreclosure, 20–21, 56–72
  markets, 59, 63
multicultural, 158, 161, 164, 166

narrative, 109
Nashville, Tennessee, 134, 157, 159–168
naturalize, 96
nature, 206, 217
neigbourhood, 20, 23–38, 158, 159, 160
  African-American and Latino, 62
  associations, 34, 161–163
  change, 157, 160
  effects, 60, 62
  politics, 20
neighbouring, strategic, 29–37

neo-liberalization, 46
neo-Marxist, 99
network, 20, 33, 48, 101–2, 184
New Urban Politics, 45–47, 51
New York City, 91–92, 102, 104
Newark, New Jersey, 62, 67, 72

offence, quality-of-life, 91
old urban politics, 46
order, 93
outdoor media, 89

panopticon, 80
parallax, 230
  capital, 58–9, 66, 67
  view, 2, 10, 223
participatory planning, 113, 116, 120, 124, 126
party politics, 99
Peck, Jamie, 47
performative, 109
place, 24–25, 27
  bundles, 37
  capital, 186
  politics, 27
planners, 91, 112
pluralist, 45
police, 10, 28, 36, 86–87, 93, 167
police order, 93
policing, 9, 37, 82, 85–98
policy, 225
  mobility, 7
political, 34, 172, 184
  parties, 105
  party systems, 106
  strategy, 25
politics, 7, 9, 10, 28, 35, 86–87, 211
pollution, 214
population density, 213
postcolonial, 148, 149
post-democratic, 10, 133
post-industrial, 140, 187, 191, 198, 229
post-industrial city, 60,
postmodern, 133
post-structuralist, 133
power, 174

private property, 92, 96, 114
public space, 125, 166
Pudl, 97

queer theory, 140–141, 150
queering, 139

race, 72
racism, 29
radical margins, 135
Rancière, Jacques, 9–10, 27, 82, 86–88, 93, 134, 226
redlining, 59
relational, 4, 6, 20, 26, 48, 225
relational, place, 23, 37
relations, 174, 176, 181, 211, 217
representation, 100, 101
representational, 102
  space, 102
repressive, 91–93
residents, 167
resistance, 123
risk, 58, 61
Robbo, 95
rootedness, 176
Rust Belt, 175
ruthless, 57

scale, 3, 26, 32, 45, 47
  critique of, 4
  jumping, 106
  rescaling, 55, 59–60
Seattle, Washington, 113, 124–126
  Ballard neighborhood, 118–120
  Municipal Code (SMC), 118
securitization, 67
segregation, 156
self-interest, 180
setting, 17–21
sexual
  identity, 142
  politics, 148
sexuality, 136
  studies, 141
Simmel, Georg, 19, 208–213
Singapore, 136, 140–151

skatepark, 113, 116–126
Small Business Administration (SBA), 177
social
  capital, 176
  class, 72, 137, 189, 197, 200
  control, 83
  inclusion, 93, 162, 172
  justice, 113
  media, 104
  reproduction, 45
spatial, 87
  ontology, 167
splintering, 48
state, 134, 177
  power, 4, 6
stereotypes, 172
Stone, Clarence, 79
structural power, 99
struggle, 43
subprime lending, 72
  markets, 67
Sun Belt, 175
supply-side explanation, 66
surveillance, 92
Sydney, Australia, 92

taxes, 177, 179
territory, 47, 49, 51, 108
Thiollière, Michel, 104
Tonnies, Ferdinand, 18, 19
topological, 6, 9, 99
  arrangement, 4
transnational, 55
  capital, 63

urban, 50
  citizen, 19
  media, 97
  planning, 113, 165
  problematic, 58
  political ecology, 212
  politics, 1–3, 6–7, 17–18, 20, 42, 44, 55, 79, 81, 83, 113, 134, 159, 183, 185, 195, 199, 201, 223, 229
    critical, 139
    sexual, 141

urban, *cont.*
  societies, 19
  regime theory, 55, 79

volunteer, 180, 181

wilderness, 213
White, 30,

Williams, Raymond, 207
Wirth, Louis, 208–213
women-owned business, 178
Worcester, Massachusetts, 172–187
working class, 194, 201

Žižek, Slavoj, 1, 2, 10, 134, 189, 193, 201